THE DEFINITIVE GUIDE TO INTEGRATED SUPPLY CHAIN MANAGEMENT

THE DEFINITIVE GUIDE TO INTEGRATED SUPPLY CHAIN MANAGEMENT

OPTIMIZE THE INTERACTION BETWEEN SUPPLY CHAIN PROCESSES, TOOLS, AND TECHNOLOGIES

Council of Supply Chain Management Professionals

Brian J. Gibson
Joe B. Hanna
C. Clifford Defee
Haozhe Chen

Vice President, Publisher: Tim Moore
Associate Publisher and Director of Marketing: Amy Neidlinger
Executive Editor: Jeanne Glasser Levine
Consulting Editor: Chad Autry
Operations Specialist: Jodi Kemper
Cover Designer: Chuti Prasertsith
Managing Editor: Kristy Hart
Project Editor: Deadline Driven Publishing
Copy Editor: Deadline Driven Publishing
Proofreader: Deadline Driven Publishing
Indexer: Angie Martin
Compositor: Bronkella Publishing
Manufacturing Buyer: Dan Uhrig

© 2014 by Council of Supply Chain Management Professionals
Published by Pearson Education
Upper Saddle River, New Jersey 07458

For information about buying this title in bulk quantities, or for special sales opportunities (which may include electronic versions; custom cover designs; and content particular to your business, training goals, marketing focus, or branding interests), please contact our corporate sales department at corpsales@pearsoned.com or (800) 382-3419.

For government sales inquiries, please contact governmentsales@pearsoned.com.

For questions about sales outside the U.S., please contact international@pearsoned.com.

Company and product names mentioned herein are the trademarks or registered trademarks of their respective owners.

All rights reserved. No part of this book may be reproduced, in any form or by any means, without permission in writing from the publisher.

Printed in the United States of America

First Printing December 2013

ISBN-10: 0-13-345392-8
ISBN-13: 978-0-13-345392-8

Pearson Education LTD.
Pearson Education Australia PTY, Limited.
Pearson Education Singapore, Pte. Ltd.
Pearson Education Asia, Ltd.
Pearson Education Canada, Ltd.
Pearson Educación de Mexico, S.A. de C.V.
Pearson Education—Japan
Pearson Education Malaysia, Pte. Ltd.

Library of Congress Control Number 2013952810

The authors would like to dedicate this book to our families. We greatly appreciate the encouragement, patience, and support that they have provided throughout our careers and this endeavor. Brian Gibson would like to thank his wife Marcia and son Andy. Joe Hanna would like to thank his wife Amy and children Mark and Sara. Cliff Defee would like to thank his wife Joi and kids Mason, Collin, Jesse, Casey, Drew, and Scott. Haozhe Chen would like to thank his wife Jing and daughter Connie.

CONTENTS

Foreword xi

1 Defining the Supply Chain 1

2 Supply Chain Structure, Processes, and Trade-Offs 29

3 Key Strategic Principles 71

4 Supply Chain Information and Technology 111

5 Managing the Global Chain 147

6 World Class Supply Chain Performance 177

7 The Supply Chain of the Future 205

Index 221

ABOUT THE AUTHORS

Brian J. Gibson, Ph.D., holds the Wilson Family Professorship in Supply Chain Management at Auburn University. Drawing upon his experiences as a logistics manager, Dr. Gibson is recognized as an innovative educator who brings a practical perspective to the classroom. He is coauthor of more than 100 academic and practitioner-focused articles, the annual State of the Retail Supply Chain Report, the Supply Chain Essentials video series, and two market-leading textbooks. Dr. Gibson is active in the SCM community, serving on the CSCMP Professional Certification Committee, the NASSTRAC Education Committee, the RILA Logistics Steering Committee, and the Accenture Academy SCM content development team.

Joe B. Hanna, Ph.D., serves as Associate Dean and Regions Bank Professor of Supply Chain Management at Auburn University. Dr. Hanna has published over sixty articles and is active in Auburn's undergraduate, graduate, and executive education programs. Joe has also performed education and/or consulting for several firms including Pratt Whitney, Hyundai, Menlo Logistics, Pacer International, Transplace, Fortna Supply Chain Consulting, SMC^3, Kane is Able, and others. Prior to entering academia, Dr. Hanna gained practical experience with Phillips Petroleum (now ConocoPhillips), Phillips 66 Chemical Company (now ChevronPhillips Chemical Company), and Coopers and Lybrand (now Pricewaterhouse Coppers).

Cliff Defee, Ph.D., is the EBSCO Associate Professor of Supply Chain Management at Auburn University. Dr. Defee's work has been nationally recognized, most recently as coauthor of articles that received best paper of the year distinctions in Journal of Business Logistics (2009) and Transportation Journal (2010). He is coauthor of the annual State of the Retail Supply Chain Report. Dr. Defee co-founded outsourced services provider PFSweb and he served as Chief Operating Officer for six years following a 13-year career with Accenture.

Haozhe Chen, Ph.D., is an Associate Professor of Marketing and Supply Chain Management in the College of Business at East Carolina University. He has published more than 20 articles in leading logistics and supply chain management journals, and his current research interests include reverse logistics, supply chain integration, supply chain relationships, and international logistics. He is especially interested in China-related logistics topics and issues. He has eight years' industry experience in international trade business in China, and he is an active member of ASTL, CSCMP, ISM, RLA, and RLSC.

ACKNOWLEDGMENTS

The authors would like to acknowledge the contributions of our mentors and colleagues for helping us develop the knowledge and skills needed to author this book. We would also like to recognize the team of dedicated supply chain management professionals who recognized the critical need for the SCPro Certification program. Rick Blasgen, Kathleen Hedland, Ann Neumann, Kathy McInerney, Heather Morys, and everyone at CSCMP. We thank Dr. Ted Stank (University of Tennessee), Dr. Chris Moberg (Ohio University), and Dr. Tom Speh (Miami University) who deserve special recognition for leading SCPro from concept to reality (and for allowing Brian Gibson to ride along on the journey). It is also important to acknowledge the essential contributions of Dr. Chad Autry (University of Tennessee) and Jeanne Levine at FT Press for creating the supply chain book series and for being patiently persistent during the development process.

FOREWORD

Supply chain and supply chain management have become household phrases supported by newspapers, television, and electronic media advertising of many companies over the last few years. Although it has become a well-known term, I don't think it has a well understood, broadly accepted definition that everyone understands, and it has not penetrated very deeply throughout the general public. One may wonder why that is. Certainly many think they know the definition of well-known business activities such as marketing, sales, and finance. Marketing, for example, while under its present concept started sometime in the eighteenth and nineteenth centuries during the Industrial Revolution. Even so, today marketing has, based on some recent articles, more than 72 different definitions authored by respected organizations and senior business leaders. In addition, most definitions given by the general public of any of the above would just be scratching the surface of each discipline, because of the evolution of supply capabilities and market saturation and evolving technologies, causing the discipline to continue to branch out into many detailed business specialties.

Supply chain management had its origins first within the marketing discipline as distribution. Later, in accumulated activities, it came under logistics, and then into a much more inclusive set of activities that were defined within this program. Supply chain management by its accepted definition includes multiple specialties as well as multiple orders of complexity. At its core, it addresses the attainment of excellence in a number of functional activities historically defined under the discipline of logistics. It then addresses the requirements of additional traditionally separate functions, requiring coordination and collaboration among themselves as well as with the balance of needed functions to operate any company. Finally, it requires proficiency to extend coordination and collaboration of the company with strategic supply chain partner organizations and many of their internal functional groups of key material and service suppliers and customers/consumers, which the company believes necessary to optimize its profit growth and level of service. I will unpack some of what I have just said to hopefully impart why beginning a learning journey using the CSCMP Pro Level program is so critical today.

But first let me state that what you will not find in this program is a cookie cutter design of a supply chain or supply chain management. In my experience, the prerequisite for any supply chain model or supply chain management organization is that it must meet the basic characteristics of the business and industry within which it is operating and follow the strategies intended for that specific business and culture and be guided by the historical disposition of the company. In addition, do not expect to find many companies having the same definition and responsibilities for even a given function within a supply chain organization. In my 40 years working within the supply chain area, I have lead a number of logistics or supply chain organizations, each of which had very different responsibilities.

Leading the logistics area in one consumer products company meant being responsible regionally for customer service, order management, finished goods distribution centers, finished goods transportation, accounts receivable and collections, information technology and sales services. Sales forecasting, order billing, and systems integration were done centrally by another area.

In another consumer products company, leading the logistics area meant being responsible nationally for sales forecasting, demand planning, order management, production planning, production scheduling, engineering, all distribution centers (finished goods, raw materials, and supplies), all transportation, and procurement.

In a healthcare manufacturing company, I lead a supply chain organization actually running as a separate company division supporting four product divisions. My organization included national sales forecasting, demand planning, production planning, customer service, order management, accounts receivable and collection, all distribution center locations, engineering, all procurement and travel services, quality management, product regulatory oversight, and information technology for the entire U.S. company and as ownership of all U.S finished goods. In addition I was the matrix global supply chain leader having both direct and indirect accountabilities for international operations and supply chain human talent. Acting as division president with an organization this large, I had my own chief financial officer and human resource officer as well.

My final example is a vertically integrated fashion retailer who sourced its designs in thousands of factories throughout the world and directly sold their products under different brands in over 6,500 retail stores and through direct catalog and the Internet mostly, but not only, in the United States. Under the parent corporation, supply chain services were a separate for-profit corporation. The areas of global responsibility included coordination of demand plans; all distribution and assembly center planning and operations; design and construction of all non-store facilities; all nonmerchandise and travel procurement; product regulatory responsibility; legal and company compliance for product technical quality; safety and country compliance for international manufacturing and importing of products for sale in the U.S.; supply chain research and development; engineering and time-definite air, ocean, train, and truck inbound and outbound transportation to 6,500 retail locations; and parcel deliveries to millions of consumers and for all nonmerchandise purchased (that is, at one time, more than 50 percent of the product left factories in Asia by air on Friday, was delivered in the U.S. disassembled, sorted, reassembled, packed, shipped, and delivered to thousands of stores across the U.S. by the following Friday in predetermined two-hour delivery windows). With an organization of this size, I also had my own financial team, human resource team, and information technology, each headed up by an executive. Finally, this company performed all of or some of these activities/services for outside third-party retail customers (some competitors or spin outs) for profit to defray some of the needed committed capital and size leveraging overall capability and cost structure for our parent corporation.

All are different experiences illustrating the different definitions each of these firms has for logistics or supply chain services. In each of the experiences, the original definition of either logistics or supply chain started out as a narrow set of activities and grew as performance was achieved and an additional break-through was needed by each company or parent company. Each evolved within the boundaries of what the company leadership could accept as a new definition of span of control for supply chain management and their inter-organizational interfaces. In each of the cases, the organizations grew not because I might have wanted them to but rather because they better fit the strategy, purpose, and capability needed for each company to better succeed and because the organization had demonstrated significant positive results against the goals of the firm and broader business skills and capabilities within the areas we managed that could take on larger business processes.

As illustrated, the strategies, business purpose, and business characteristics when I was part of a healthcare company were dramatically different from those required by the fashion retailer later in my career. Both had products to sell, but the business economics—hence the speed, cost and agility of the supply chains—were vastly different, each designed to meet its own strategic and financial purpose and intent. For both companies, the profitable selling of certain products was the ultimate intent with the supply chains enabling those goals. If I had worked for UPS (United Parcel Services) or FedEx (Federal Express), my whole reason for being would have been to be a supply chain product service company as supply chain services are what they are selling. All companies have a defined or de facto business product strategy that must be enabled by a defined or de facto supply chain strategy. As you grow in your learning and experience, you should become the positive catalyst that provides insight to your company's overall strategy.

Supply chain leadership facilitates and sponsors not only the functional and process competencies of its defined areas but also across its organization and key strategic partner organizations. Being a supply chain leader requires you have achieved and maintain expertise in your area of responsibility and a good working understanding and empathy about how your role fits within the bigger picture, organization, and in achieving the overall goals of your business. Expertise is much like leadership and reputation. While it requires learning and proficiency, you are an expert only when others witness and recognize your level of demonstrated results and are willing to depend on you for advice, counsel, and direction.

The learning to be achieved from these courses will provide a framework of knowledge to understand the direct components of supply chain management that exist today in the traditional organizational models of most companies. The series of studies and certifications will directly cover the planning of business demand; the supply and processing activities needed to produce and support business products and services; the manufacturing, assembly, and/or service activities in creating salable products and services; the managing of transportation for supplies, services, components, and end products to achieve a sale and/or consumption; the activities that determine and manage the amount of resources needed over time to support suppliers; component and final product needed to support demand and service; the activity to plan and operate physical locations that receive, store, and ship supplies; components and end products to satisfy production; assembly and distribution; and finally, the fulfillment of final sale and possible aftermarket service interfacing with customers and/or consumers.

The first order of business for any supply chain associate is to understand each of the business activities individually sufficient to know how your current responsibilities meet the high bar of proficiency while complimenting the other direct supply chain activities and other functions within your company. For any of the previous activities, a valued supply chain associate should know and be able to practice with high proficiency the basic concepts of the activity and the existing role within their organization. I say existing role because supply chain management done competitively is an evolving discipline needed but not necessarily existing in many organizations today. Part of your expected role after completing this certification program is to learn how ideally the roles and collaboration should take place, to test and over time improve the roles in your area and related supply chain areas to a higher level of proficiency, supporting increased profitable growth. Your education within your area of expertise and beyond does not end with these programs. Daily new insight by others with the use of ever evolving technology will require continual reading, discussion, and experimentation if you want to stay the best and grow into more comprehensive roles.

Each of the above activities has underpinnings of key elements and processes that are based on sound principles to improve overall effectiveness and efficiency while demonstrating responsive and environmental sustainability. Each of the above activities, processes, and outcomes due to advancing technology today does, or will in the future, need to be transformed or continually improved in how it functions, in reassessing its purpose, and what can and should be known by each supply chain partner to better anticipate or eliminate unnecessary steps in achieving world class supply chain accomplishments. But to leverage today's technology, you must know each activity's reason for being, how and why we carry it out the way we do today to be able to replace and/or enhance it through the use of new technology to transform and dramatically improve your department or your company's business outcomes. Let me give you a few examples from my past that impacted the companies I worked for, were transformative to traditional skills, and came about because of competent functional team members willing to get out of their respective responsibility boxes and think more broadly to improve business outcomes.

While employed by a consumer products company, I needed to reduce capital and expenses while improving our service to our customers. At the time, all of our products were shipped to customers under tight, time-definite deliveries from our distribution centers, which were supported by shipments from our manufacturing plants. By bringing together a cross functional team, we were able to create a new customer offering of plant direct shipments for some of our products at better sales prices within the same delivery window requirements while bringing down our cost of storage and transportation. It happened by all the functional areas of order management, distribution supply, production scheduling, and sales being willing to share responsibilities and information and have joint direction and changes incorporated within the information systems. The changes included filtering potential customer orders (50–60 percent of plant-made products) back to our plants for direct shipment at better customer prices. Once changed to direct plant shipments, computer systems would automatically reroute the customer order with required carriers and transit times to meet delivery requirements, adjust original regional distribution center and plant shipment forecast balances by time bucket, by location, and by product, reduce pending distribution center replenishment quantities, and divert inventory requirements to the customer plant direct order or commit time-definite planned production to that plant customer shipment. I don't have to tell you that trying to do any one of these procedural changes individually would have been highly unlikely and singularly having each required system change pushed through as a high priority information technology change would probably be impossible. But, together, the ideas came forward. Together, they were discussed until they became a new and better collaborative plan for the business and together the group got sufficient support to convince leadership of the needed system, process changes and expected results, and approval and implementation.

Another I believe very good example of a cross-functional breakthrough came when I was employed by a healthcare manufacturer. We were selling medical supplies through distributors who then resold them in smaller quantities to hospitals, clinics, group physicians' offices, individual offices. The forecasting process for the business did not perform very well and was a continual challenge to both manufacturing and logistics areas in that the manufacturing process due to high set-up costs required long runs followed by sterilization processes which were either location or capacity bound due to known allowable technologies. Poor forecast results with required service levels in the critical health field required high inventories and costly manufacturing changes. By bringing together a cross-functional team, we found that the company rebate billing systems for the distributors over the years had been modified to require distributor customers to daily electronically remit their shipment information delineated by each of their individual end-user customers. Simply the use of this information could tell us when and how much of each product was being shipped to the final "retail" use location, hospital, clinic, physician, etc. The company in its billing area knew daily what we shipped to each distributor and what each distributor shipped to each end user point. Using this information for supply chain purposes, we had near full consumption level pipeline visibility of our products as well as the ability to reasonably determine daily inventory balances at all distributors. By using this information for supply chain purposes, the team was able to make a much more accurate short-term demand plan, distribution requirements, and manufacturing plan, which improved the predictability of what products needed to be allocated and shipped to each key customer and when to project our ultimate use of products by or for patients. They made the supply chain process materially better not by better estimates and statistics but by better, more timely, more deeply defined demand data. In today's world, the phrase that best describes this capability would be using *Big Data* in a supply chain planning and execution process. Although the information was not real-time, it measurably provided more comprehensive and timely status of what was actually going on daily with our end-users and our distributors and allowed us to better plan at a lower level of product inventory than ever before while achieving our service goals.

Again, bringing together supply chain-related talent—going back to the primal purpose as to why we do our work and continually rethink what kind of information and/or process change might be available today that could transform our results faster in a more positive manner—requires experts in all fields who are willing to collaborate for a joint and better idea. It is also for every supply chain associate the unconditional requirement to understand your areas of scope or responsibility soundly and those areas you interface with sufficiently to be able to develop new provocative ways to accomplish better solutions and results. All the previous examples reinforce to me and I hope to you how critical taking and completing this program is to your future success.

While learning more about each of the previous direct components of supply chain is important, your learning agenda will not be complete. You will have to add to the knowledge of this program a general understanding of the other business functions that make up any complete company to be able to both appreciate their interactions within any company model and to also help make better decisions by you, your direct reports, and ultimately your company. For example, understanding the specific direct and indirect impact your area's activities have within the financial performance of your company is important. Being able to translate your area actions into known business levers that are better understood by all in the company elevates the critical nature of your or your group's efforts—financial acumen is a basic skill needed for your success. Become knowledgeable in your company's profit and loss statements, balance sheet, and cash flow statements. If you are not familiar where your role is represented in these reports, seek advice from your finance department. If you're unfamiliar with a company's financial relationships, research through Google "DuPont Model" and you will find definitions and explanations to decouple the activity of any company reflected in financial information providing insight into how business assets, expenses, and working capital are levers to describe a company's performance. Drill down deep enough within your company's reports to find out where the assets you use show up, where all your direct expenses supporting your business activity show up, and where the expenses you manage show up (that is, transportation expenses show up in all physical items of the company from raw materials to finished products, for all promotional material, building, and office materials, or any other physical supply you company uses including paper and pencils).

You should be able to start accessing what proficiency you can bring to any of the previous to improve the business results of your company. You may also find the need to add activity-based cost management for supply chain-related services to properly report the effort and costs to each brand or product line to more accurately charge and show which brands or product lines or their practices are consuming the resources of your supply chain department or organization and absorbing the true respective costs. If you have never done this, it will certainly open a new window of knowledge for you and your team and may open some new profit improvement programs for your company.

Whether you are talking about one or all of the components of supply chain management, or all the activities of your company, you must be able to understand and work within the current culture of your company while creating and maintaining increasing circles of productive relationships. As you learn new and better methods to do your work from a program like this, work with others or see how you and others might need to change your approach to work that effects both of your areas. You will need to think about how to successfully introduce change to your team, to your peers, or to your superiors. One of my early mentors advised me that I should think about my plans to improve things like one thinks about a sailing ship heading for a new destination. He advised that the wind of resistance for the status quo will continually attempt to take my efforts off course and I should be prepared like a good sailing ship to continually tack my effort of change to compensate for the winds of resistance. As I dealt throughout my career with the many times I needed to change the behavior, skill, and culture of an organization I joined, I appreciated his guidance. Supply chain leaders by the very nature of the work managed tend to be more analytical and at times can miss the nuances so important in working with your own and other functions. My recommendation is to find a way to directly "step into the shoes" of the other functions within a company. Experience by mentoring or observing the challenges of selling to a customer or placing calculated bets on the volume of planned business down to the specific product meeting. Continually changing demands on manufacturing sites is an area in which you should try to get some experience or at least observation of management processes. None of the other activities or functions within a company is any easier to manage; they are only different. Not having some knowledge of how and why other company leaders are driven within your company leaves your best proposal at a disadvantage.

Over my lifetime I have enjoyed working for a number of great companies in multiple industries. One of the techniques I chose early in my career was to find a way to get a better understanding of the company I worked for, what its objectives were over time and how it performed against the objectives; how the culture evolved over time; and what their practiced behaviors were. My unusual first request each time I joined a company was to be able to review the last five years of the company's strategic plans if they existed. If strategic plans were not available, I asked for five years of the company's total budget plans to whatever level was available. Most times, I was able to secure either of these documents I have to say with a very inquisitive look. Close review of either of these documents told me a lot about my new company, new employee associates, what the company planned for annually, and what they were able to achieve and why, what they put great emphasis on and what was left as unsaid or in some cases not related to the overall purpose of the company. Doing this gave me a starting point to understand my role and fit in the company. It many times gave me the understanding and insight to be able to change my role or the role of supply chain management within the company. Later in my career, it gave me the ability to create and present strategic supply chain strategies that dovetailed into the overall purpose and business plan that focused on producing a more competitive and profitable growth for my companies.

As you grow through your career, I believe you will find that to continually excel, you will constantly have to improve your functional skills and human operating skills and then for the people you lead. Your life learning agenda will be supported by programs such as this but also must be added to with a heavy dose of human operating skills. By that, I mean skills that teach you and eventually your team the principles that are fundamental to effective leadership. These principles include how to better understand yourself and your style, how to deal and influence others, how to manage change, how to feel more empowered, and over time, how to better coach individuals, shape a culture, and eventually get more results in less time. For most people (including myself), you come to learn that demonstrating high levels of competency in a position, department, division, or company is only part of what you need to have the recognized skills and accomplishments to succeed in most businesses.

When I first had an opportunity to create a separate free standing supply chain organization for a healthcare firm, it happened first because senior management had developed an increasing awareness of supply chain performance and there was continuing pressure to reduce the total cost of medical supplies. We thought if we could focus on both the company's and the industries supply chain inefficiencies, increase the acceptance of new levels of standards, and increase capabilities identified using enabling technologies, we could materially reduce total cost to the patient and increase our growth profitably. To accomplish this mission, the company needed a new supply chain management capability providing a focal point for all supply chain processes, all product businesses, and all local market customers and channels. The new capability needed to promote supply chain integration, creating economic value throughout the channel by providing higher quality fulfillment processes and service levels, streamlining all internal and external interfaces, and marketing value of these capabilities through basic and advanced offerings.

Changes to supply chain management were only a part of the cultural transformation that was going to take place globally across other key areas including geography, technology, sales, marketing management, and human resources. What the entire organization needed to learn was the skill of change management— the continuous process of aligning an organization with the marketplace and doing it more responsively and effectively than competitors. We needed to know the most contemporary thinking in all functions and processes of supply chain to help define a new and better business solution for the company and introduce progressive thinking and influence to our industry. We needed to acquire new human operating skills that redefined or refocused our organizational and relationship abilities. This included honing our listening skills to understand new potential and possibilities, becoming more comfortable with the new bigger company teams being developed from all business backgrounds, and understanding each of our own personal filters that direct us to our own selective perception and that can create unconscious blind spots. We needed to put in the same effort to understand how each of us was affected by our own thought habits and behavior styles. We all needed to understand what each of us could do more and how every one of us could better support each other all to get better results. All the previous, if done well, will be a positive culture-shaping transformation while better enabling competent functional and/or process leaders to create a better function, department, supply chain organization and company. I mention this experience because it profoundly influenced what I needed to know to be more successful and satisfied both in my professional and personal life.

Two quotes on certificates received by my team and me from the senior management of the healthcare company summarize better what I am suggesting in successfully completing this program and augmenting it with further learning and experience within this discipline and beyond to become a successful supply chain professional:

> "For extraordinary work on developing recommendations for transitioning processes, organization alignment and internal relationships from current Corporate, Hub, and Divisional structure to one which includes a new operating service division to achieve service leadership for our company."

> "For recognition to the founding supply chain services operating committee for creating a new one company division that through its leadership transformed the division to a more empowered culture while delivering world class service and financial benefits recognized by our distributors, end-user customers, and product businesses."

In taking this program, you will develop a broader understanding of supply chain management principles, which will be a great foundation to help you build out a more effective leadership role that will require a lifetime commitment to advancing supply chain and human operating teachings and practices.

Also remember while going through this program that in the beginning, supply chain management as a discipline was burdened by a lack of timely integrated information technology to support it. Functional areas, however, continued to build out their individual expertise and became better skilled in creating and maintaining productive relationships with the logistics-related areas around them and the necessary other business functions they needed to work with to do their responsibilities as best as they could. The result of this effort much improved business outcomes but still had many organizations push their way to a better supply chain and channel outcomes. I believe that most companies, if they had their choice, would want to have a *sense- and response-based* supply chain capable of pulling whatever it needed through its business and its suppliers to meet and achieve the product and service goals it set out for itself. Up until now, that has not been completely possible. But with the capability of information technology today, I believe we could be at the point of crossing the Rubicon, of demonstrating throughout many industries the ability to operate newly designed sense and response supply chains by the use of Big Data that virtually and physically pulls product or service needs through supply chains and channels to achieve service-based or end-user consumption. It is the ability of having, sharing, and properly using ubiquitous status information across and among a complete business supply chain that will be the transformative capability. This will not change the fundamentals you are about to learn, but rather give the profession the ability to use them differently to achieve a high order of results or solutions in less time with less assets or expenses deployed more consistently and profitably. All of the components of this program are building blocks to a foundation of understanding many in most companies do not have, never thought they needed but that will be required to support business transformations in the near future. Building off of this platform makes you a better well-rounded informed employee, which, with demonstrated business solutions and results can be a prime candidate for higher leadership positions in almost any well-run company.

Each time I started a new supply chain organization, I reminded myself and my team that the journey of change management needed to begin with us to achieve supply chain service leadership.

So, to you I say, "The journey to twenty-first century supply chain management begins with you. Learn, explore, execute, excel!"

—Nicholas J. LaHowchic
CEO & President of Diannic, LLC
Coauthor *Start Pulling Your Chain! Leading Responsive Supply Chain Transformations*
Former Executive Vice President of Limited Brands, Inc., former CEO & President of Limited Logistics Services, Inc., and past President of Supply Chain Services, Becton Dickinson

1

DEFINING THE SUPPLY CHAIN

Supply chain management is a vital, yet often underappreciated facilitator of trade that fosters customer convenience, business success, and societal development. Consumers benefit greatly from supply chain management, yet few people think about the planning, cost, or activities involved in getting fuel to the filling station, fresh foods to the store shelf, or essential medical supplies to the hospital emergency room. People just assume that products will be readily available without worrying about how much their quality of life depends on productive, efficient supply chains.

The same situation exists within many organizations. Despite the ability of supply chain management to facilitate cost-savings and a competitive advantage, relatively few individuals in marketing, finance, or manufacturing pay much attention to it. They primarily think of supply chain management in terms of operational activities that occur behind the scenes to complete a customer transaction. The only time that these individuals focus on the supply chain is when a supply disruption, manufacturing shutdown, or delivery delay occurs. They expect supply chain managers to quickly resolve the issue, return the organization's supply chain to a state of balance, and take steps to prevent future occurrences.

The good news is that savvy organizations such as Amazon.com, McDonald's, and Unilever recognize the importance of supply chain management and make it a strategic priority. They understand that it is impossible to compete effectively in isolation of their suppliers, customers, and other entities in the supply chain (Lummus & Vokura, 1999). This is critical in a complex global economy where your suppliers and customers may be on different continents, omnichannel fulfillment capabilities are needed, and service expectations are rising. Taking the time to develop efficient and agile supply chain capabilities to respond to these dynamic market requirements is the difference between great success and utter chaos.

The first step in the journey to supply chain management success is to understand its foundation concepts. A discussion of what supply chain management is, why it is important, and how it benefits the organization is needed to get everyone on the same page for the pursuit of supply chain excellence. Hence, this chapter provides a level-setting discussion of key terminology and definitions. The focus then turns to the purpose and goals of supply chain management to clarify the essential objectives that supply chain managers must pursue. Next, a review of the evolution of supply chain management and its key participants is provided. The chapter closes with an introduction to the value proposition of supply chain management and its capability to drive organizations toward better, faster, and cheaper demand fulfillment.

Key Terminology and Concepts

A fundamental challenge in supply chain management is the lack of a common "language" that is used across organizations and industries. Unlike financial accounting and other long-established business fields, there is not yet a universally accepted set of definitions and rules that drive supply chain management. For example, asking a group of business executives to simply define the term *supply chain* would lead to a long and potentially contentious discussion.

Such a situation is not unusual in a relatively new field like supply chain management. Initially, there will be a lack of consensus as to its definition or consistency in its application (Ballou, Gilbert, & Mukerjee, 1999). While this is to be expected, it is not desirable. Consistent definitions are essential for understanding the basic characteristics and scope of supply chain management. They provide clarity regarding what supply chain management is and is not, drive acceptance of its key elements, and facilitate its application (Gibson, Mentzer, & Cook, 2005).

This section reviews the popular definitions of essential supply chain management terms, evaluates their common components, and highlights the scope of the field. Having this solid frame of reference will help you avoid the dangers of defining the field too narrowly or too broadly. A narrow perspective will limit the potential value of supply chain management to your organization. In contrast, an overly broad conceptualization will make it difficult to establish control over the processes, foster collaboration, and control performance.

Supply Chain Concepts

Before an organization tries to focus on supply chain management, its leaders must determine what the supply chain encompasses. Just as you can't manage what you don't measure, you can't plan and execute what you haven't clearly defined. Hence, it is important to articulate the overall purpose, scope, and components of a supply chain. Following are useful supply chain definitions that highlight critical aspects of a supply chain.

- **From the Council of Supply Chain Management Professionals (2010)**—The material and informational interchanges in the logistical process, stretching from acquisition of raw materials to delivery of finished products to the end user. All vendors, service providers, and customers are links in the supply chain.

- **From Christopher Martin L. (1992)**—The network of organizations that are involved, through upstream and downstream linkages, in the different processes and activities that produce value in the form of products and services delivered to the ultimate consumer.

- **From Coyle, Langley, Novak, and Gibson (2013)**—A series of integrated enterprises that must share information and coordinate physical execution to ensure a smooth, integrated flow of goods, services, information, and cash through the pipeline.

One important feature of these definitions is the concept of an integrated network or system. A simplistic depiction of a supply chain, as featured in Figure 1-1, suggests that a supply chain is linear with organizations linked only to their immediate upstream suppliers and downstream customers. It also focuses on only one-way material flow, which fails to consider vital information and financial flows, as well as reverse material flows. Such misconceptions oversimplify reality and fail to reveal the dynamic nature of a supply chain network.

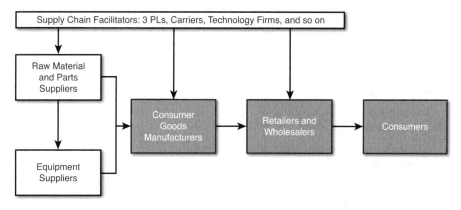

Figure 1-1   Linear representation of a supply line

In truth, supply chains require a multiplicity of relationships and numerous paths through which products and information travel. This is better reflected by the conceptual diagram of a supply chain in Figure 1-2, in which the supply chain is a web or network of participants and resources. To gain maximum benefit from the supply chain, a company must dynamically draw upon its available internal capabilities and the external resources of its supply chain network to fulfill customer requirements. This network of organizations, their facilities, and transportation linkages facilitate the procurement of materials, transformation of materials into desired products, and distribution of the products to customers.

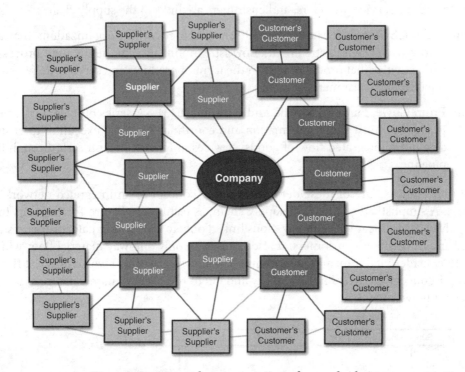

Figure 1-2 Network representation of a supply chain

Simple representations aside, it is critical to understand that no two supply chains are exactly alike. An organization's supply chain structure and relationships will be influenced by its industry, geographic scope of activity, supply base, product variety, fulfillment methods, and demand patterns. Consider, for example, a multinational manufacturer and a local farm-to-table restaurant. Both organizations would benefit from strong and stable supply chains. However, the manufacturer's network is at greater risk of disruption and must integrate geographically diverse suppliers with multiple selling channels.

Supply Chain Management Perspectives

Introduced in the early 1980s, the term *supply chain management* began to take hold in the mid-1990s and is now part of the everyday business lexicon. Whereas a supply chain is an entity that exists for the fulfillment of customer demand, supply chain management involves overt managerial efforts by the organizations within the supply chain to achieve results (Mentzer et al., 2001). These efforts can be strategic or operational in nature, though the vast majority of respondents to a Council of Supply Chain Management Professionals survey indicate that the primary role of supply chain management within an organization is a combination of strategy and activity (Gibson, Mentzer, & Cook, 2005).

Defining supply chain management would seem to be a straightforward task, yet it has been a vexing challenge with the introduction of many alternatives. A Google search for "supply chain management definition" quickly yields about 12,000 results. Among this plethora of descriptions, you will find professional associations, consultants, and academicians addressing similar issues but providing their own interpretations and areas of emphasis. Following is a sampling of relevant definitions:

- **From the Council of Supply Chain Management Professionals (2011)**—The planning and management of all activities involved in sourcing and procurement, conversion, and all logistics management activities. More important, it also includes coordination and collaboration with channel partners, which can be suppliers, intermediaries, third-party service providers, and customers. In essence, supply chain management integrates supply and demand management within and across companies.

- **From Gartner (2013b)**—The processes of creating and fulfilling demands for goods and services. It encompasses a trading partner community engaged in the common goal of satisfying end customers.

- **From LaLonde (1997)**—The delivery of enhanced customer and economic value through synchronized management of the flow of physical goods and associated information from sourcing to consumption.

- **From Stock and Boyer (2009)**—The management of a network of relationships within a firm and between interdependent organizations and business units consisting of material suppliers, purchasing, production facilities, logistics, marketing, and related systems that facilitate the forward and reverse flow of materials, services, finances, and information from the original producer to the final customer with the benefits of adding value, maximizing profitability through efficiencies, and achieving customer satisfaction.

Although the definitions vary in length and complexity, they collectively focus on three themes: activities, participants, and benefits (Stock & Boyer, 2009). That is, organizations must plan and coordinate supply chain activities among their network of suppliers and customers to ensure that the end product is available to fulfill demand in a timely, safe, and cost-efficient manner. When this is accomplished, the benefits of enhanced customer satisfaction and retention will be achieved.

Related Terms and Concepts

Supply chain management encompasses a number of business processes, activities, and goals that are discussed throughout this book. Before moving forward, it is valuable to clarify their meanings and relevance to supply chain management.

Logistics Management

Logistics is a fundamental set of supply chain processes that facilitates fulfillment of demand. The goal is to supply the right product or service, at the right place, at the right time. The Council of Supply Chain Management defines logistics management as "that part of supply chain management that plans, implements, and controls the efficient, effective forward and reverse flow and storage of goods, services and related information between the point of origin and the point of consumption in order to meet customers' requirements." Whether provided internally, by a supplier, by the customer, or by an external logistics services provider, these capabilities are essential for achieving supply chain success.

Supply Management

Supply management focuses on the identification, acquisition, access, positioning, management of resources, and related capabilities the organization needs or potentially needs in the attainment of its strategic objectives (Institute for Supply Management, 2010). For most organizations, logistics controls the distribution of products; whereas supply management controls the strategic sourcing of direct materials, finished goods, services, capital equipment, and indirect materials. Both are needed to ensure optimal performance of the supply chain.

Value Chain

The concept of a value chain was developed as a tool for competitive analysis and strategy. It is composed of primary activities (inbound logistics, operations, outbound logistics, marketing and sales, and service) and support activities (infrastructure, human resource management, technology development, and procurement) that work together to provide value to customers and generate profits for the organization (Porter, 1985). A value chain and a supply chain are complementary views of an extended enterprise, with integrated supply chain processes enabling the flows of products and services in one direction, and the value chain generating demand and cash flows from customers (Ramsey, 2005).

Distribution Channel

Distribution channels support the flow of goods and services from the manufacturer to the final user or consumer (Council of Supply Chain Management Professionals, 2010). An organization can establish direct channels to consumers or rely upon traditional intermediaries such as wholesalers and retailers to facilitate transactions with final users. The rapid expansion of the Internet as a key selling platform is forcing manufacturers and retailers to develop innovative and flexible "omnichannel" capabilities in their supply chains to fulfill customer demand from stores, distribution centers, and production locations.

Key Participants

The definitions of supply chain management allude to a wide variety of entities that participate in the two-way flow of materials, information, and money. The participant network varies in size and scope, depending on the products involved, geographic dispersion of supply and demand, and customer service requirements. It is safe to say that no two supply chains are exactly alike, and a participant's role may vary in each network.

Compare, for example, the supply chains for apples versus Apple iPads. If you purchase a home with a small apple orchard on the property, you could open up a storefront to sell the apples. This simple supply chain has two primary participants: a retailer (you) and consumers. In contrast, Apple relies upon a global network of component suppliers to make key parts for the iPad, a contract manufacturer to assemble the product, transportation and logistics companies to distribute the product to global markets, retailers to sell the products, and end consumers to buy the iPads. Other organizations supplement the network with needed information, packaging, credit, and services. This complex network can be difficult to manage and costly to execute.

A logical segmentation basis for supply chain participants is their ownership stake in the product. Entities that own the goods at various stages of the supply chain are direct stakeholders. This group includes the final consumers or end users of the goods, retailers, distributors, manufacturers, and suppliers. Entities that support the flow of materials, information, and money are supply chain facilitators. They do not typically take title to the goods but play a critical role in the safe, efficient execution of supply chain activities. This facilitator group includes logistics services providers, information technology companies, consultancies, financial institutions, government agencies, equipment providers, and indirect materials suppliers.

Direct Stakeholders

Although every participant can affect supply chain performance, no other direct stakeholder is as important as the end user of the goods. End user demand is the catalyst of all activity in the supply chain, but if no demand exists, there is no need for the supply chain network. That is why so many supply chains focus on end user demand to drive planning and activity. In a consumer product supply chain, the end user is the retail consumer. In an industrial setting, the end user is a company that buys materials, goods, and services to support its operations. Examples include Lufthansa buying 747 jets and repair parts from Boeing and UPS buying diesel fuel for its fleet.

Retailers play a critical role in the supply chain, acting as intermediary between end consumers and product manufacturers. Retailers accumulate inventory from multiple sources to assemble a wide assortment of products for sale. For example, a Wal-Mart Supercenter has more than 100,000 different items in the store. In addition, retailers provide consumers with convenient one-stop shopping, competitive pricing, and financial transaction services. Retailers provide manufacturers with shelf space for their product and visibility of demand from point-of-sale data.

Wholesalers and distributors are intermediaries that provide value added services to manufacturers and retailers. Wholesalers buy products in bulk from manufacturers and sell the products in smaller quantities to retailers, provide storage facilities to reduce the need for manufacturers and retailers to hold large inventories, and offer delivery services to retailers. Similarly, distributors provide fulfillment efficiency as middlemen between manufacturers and industrial buyers. The distributor buys large quantities of materials or parts from a manufacturer and then creates smaller selling units and fulfills orders to end users in a timely fashion. This allows the manufacturer to focus on production and larger deliveries to distributors rather than managing small quantity orders from a global customer base.

Manufacturers provide the form utility of goods by transforming raw materials, parts, and components into products that are beneficial to end users. The transformation process can be completed in-house or outsourced to a contract manufacturer. The latter group builds products under the brand or label of another firm. For example, Nike product designs and specifications are produced by contract manufacturers in 777 factories in 43 countries around the world (Nike, Inc., 2013). The production processes used by a manufacturer—build to stock, configure to order, and engineer to order—has a significant influence on the design and operation of the supply chain.

Suppliers include a wide array of supply chain participants that provide essential inputs to the production process of a manufacturer. This broad category of organizations includes raw material extractors and processors, parts producers, component assemblers, and similar entities that support the creation of finished goods. Tier 1 suppliers feed critical items directly to the manufacturer; Tier 2 and Tier 3 suppliers support their downstream counterparts with a steady supply of needed materials. Suppliers bring a level of expertise and efficiency to the supply chain that few manufacturers could generate on their own.

Facilitators

The vast majority of supply chains depend upon logistics services providers to plan and execute the flow of goods from origin points to destinations. Their capabilities include inventory management, transportation, storage, order fulfillment, and related functions. Some logistics service providers focus on a single activity such as truckload transportation; others offer an integrated set of logistical capabilities for customers. These organizations invest in the equipment, talent, technology, and facilities needed to provide exceptional service. Customers can leverage these capabilities on an as-needed, variable cost basis.

Technology firms facilitate rapid flows of critical information across the supply chain. Rather than developing software in-house and trying to integrate it with other systems, direct stakeholders rely upon technology companies to provide supply chain planning, execution, and event management tools that generate cross-chain visibility, increase control, and support decision making. Some technology firms focus on specific solutions; others provide integrated suites of supply chain software.

Indirect material suppliers provide goods that support the operation of the supply chain, but are not directly associated with a specific product. These include consumables, tools, and supplies that facilitate the efficient production of goods. Similarly, packaging and material handling supplies are needed to ensure the safe and accurate delivery of goods.

Financial institutions and government agencies also have important roles in the supply chain. Banks and related institutions facilitate trade through working capital management, payment and cash management, and contract execution support. They help to reduce risk in global supply chain transactions and to reduce inventory costs. Government regulatory agencies mandate product standards, labor laws, equipment requirements, and transportation regulations to promote supply chain safety. Other agencies provide import/export support to encourage trade, control borders to ensure supply chain security, and collect fees to support the supply chain infrastructure.

Unique supply chains involved in the management of reverse flows, services, projects, events, and other unique scenarios will require the use of additional facilitators and specialists. Consultants, project managers, recycling companies, equipment manufacturers, construction companies, and laboratories are just some of the ancillary participants that support specialized supply chains.

A key to success in supply chain management is to actively engage essential participants in the planning and development of your key requirements. Information sharing about expected demand, timing issues, location, and special needs is essential for all participants. This dialogue with direct stakeholders and facilities will help them marshal the necessary capacity, inventories, and labor needed to pursue perfect fulfillment of demand.

Purpose and Goals

The definitions of supply chain management indicate that it is a complex undertaking that extends beyond the scope and capabilities of a single organization. Significant effort is needed to build and maintain a supply chain network. This involves a tremendous action list that requires expertise, time, and money—establishing strategies, building relationships and roles, aligning processes, developing people, implementing technology, and investing in capacity.

Given these requirements and challenges, it is logical to wonder whether the pursuit of supply chain management capabilities is worthwhile. The succinct answer is yes because organizations need strong supply chain capabilities to profitably compete in the marketplace. Their key goals for supply chain management should be to achieve efficient fulfillment of demand, drive outstanding customer value, enhance organizational responsiveness, build network resiliency, and facilitate financial success.

Goal 1: Achieve Efficient Fulfillment

On the most basic level, the purpose of supply chain management is to make inventory readily available in customer facing positions to fulfill demand. The fresh produce business adage "you can't sell from an empty wagon" highlights this fundamental purpose of supply chain management.

Organizations must pursue the goal of matching supply with demand in a timely fashion through the most efficient use of cross-chain resources. Supply chain partners must work together to maximize resource productivity, develop standardized processes, eliminate duplicate efforts, and minimize inventory levels. Such steps will help the organization reduce waste, drive out costs, and achieve efficiencies in the supply chain.

Reduction of supply chain expenses is a popular goal, particularly during times of economic uncertainty when companies desire to conserve capital. Efficiency initiatives can focus on any aspect of supply chain operations, though transportation and inventory are frequent cost control targets. Together, they account for 81 percent or $1.08 trillion of U.S. business logistics system costs (Council of Supply Chain Management Professionals, 2013).

Ocean Spray, an agricultural cooperative that produces fruit juices and foods, was able to cut freight costs after opening a regional distribution center in Florida. The facility reduced distances to customer locations and was well positioned to leverage empty railroad boxcars traveling from New Jersey to Florida. The shift from truck to rail, along with the reduced outbound mileage, helped Ocean Spray cut freight costs by 40 percent and carbon dioxide emissions by 20 percent (Bradley, 2013).

Kimberly-Clark, a manufacturer of personal care products, has been on a 6-year journey to create a demand-driven supply chain. The company has realigned its distribution center network and streamlined the number of facilities to take inventory and costs out of the system.

To further streamline safety stock inventories and reduce associated costs, the company is using demand planning software with retailer point-of-sale data to understand demand and develop more accurate forecasts. Over an 18-month period, Kimberly-Clark reduced its finished goods inventory by 19 percent (Cooke, 2013).

Goal 2: Drive Customer Value

Cost efficient fulfillment and inexpensive products are important, but supply chain managers must also focus on value creation for their customers. Customers are the lifeblood of the organization and create the need for a supply chain. Hence, a fundamental objective in supply chain management must be to consistently meet or exceed customer requirements.

The goal of driving customer value begins with a market-driven customer service strategy that is based on clearly understood customer requirements. Supply chain strategies, design, and capabilities should emanate from these requirements (Sweeney, 2011). The result will be higher-quality service, reduced variability, and fewer exceptions to address.

Highly consistent, just-in-time delivery is critical to the restaurants and food service companies supplied by McCain Foods, the world's largest manufacturer of French fries. Rather than focus on low–cost rail transportation, McCain works closely with a long–haul truckload carrier to provide exceptional on-time delivery performance for these time–sensitive supply chains. They preload trailers, secure additional capacity, and expedite deliveries as needed to ensure that French fries are always on the menu (Partridge, 2010).

It is important to note that Goal 1 and Goal 2 are not mutually exclusive. To succeed, organizations must establish supply chains that balance efficiency with effectiveness to optimize overall performance. The annual Supply Chain Top 25 rankings by Gartner, Inc. (2013a) identify companies that accomplish both goals by integrating demand, supply, and product into a network that that orchestrates a profitable response to ever-changing customer demands. Table 1-1 highlights the 2013 supply chain leaders based on industry opinions, 3-year weighted return on assets, inventory turns, and 3-year weighted revenue growth.

Table 1-1   The Gartner Supply Chain Top 25 for 2103

Rank	Company	Peer Opinion	Gartner Opinion	Return on Assets	Inventory Turns	Revenue Growth	Composite Score
1	Apple	3203	470	22.3%	82.7	52.5%	9.51
2	McDonald's	1197	353	15.8%	147.5	5.9%	5.87
3	Amazon.com	3115	475	1.9%	9.3	33.6%	5.86
4	Unilever	1469	522	10.5%	6.5	9.0%	5.04
5	Intel	756	515	15.6%	4.2	11.4%	4.97
6	P&G	1901	493	8.6%	5.8	3.6%	4.91
7	Cisco Systems	1167	517	8.5%	11.2	7.8%	4.67
8	Samsung Electronics	1264	298	11.6%	18.5	15.7%	4.35
9	Coca Cola Company	1779	278	11.7%	5.5	14.0%	4.33
10	Colgate-Palmolive	794	324	18.9%	5.2	3.6%	4.27
11	Dell	1409	342	6.2%	30.7	-0.6%	4.05
12	Inditex	745	221	18.0%	4.2	13.4%	3.85
13	Wal-Mart	1629	282	8.8%	8.1	4.9%	3.79
14	Nike	955	236	14.1%	4.2	10.6%	3.62
15	Starbucks	808	159	16.5%	4.8	11.5%	3.41
16	PepsiCo	810	314	8.6%	7.8	10.5%	3.41
17	H&M	399	41	28.2%	3.7	6.7%	3.22
18	Caterpillar	714	247	5.8%	2.8	23.4%	2.91
19	3M	999	105	13.3%	4.2	6.9%	2.87
20	Lenovo Group	397	211	2.5%	22.2	29.8%	2.75
21	Nestlé	679	112	13.3%	5.1	-0.6%	2.51
22	Ford Motor	552	231	5.7%	15.1	3.1%	2.51
23	Cummins	74	139	13.3%	5.3	13.5%	2.48
24	Qualcomm	122	45	12.7%	8.5	25.9%	2.37
25	Johnson & Johnson	730	144	9.6%	2.9	3.3%	2.35

Composite Score: (Peer Opinion*25%) + (Gartner Research Opinion*25%) + (ROA*25%) + (Inventory Turns*15%) + (Revenue Growth*10%)

Goal 3: Enhance Organizational Responsiveness

Another important rationale for supply chain management capabilities is responsiveness to change. The current business environment is one of rapid change with multiple forces shaping how businesses operate and survive. Supply chain management can help organizations adapt to the challenges of globalization, economic upheaval, expanding consumer expectations, and related issues (Coyle et al., 2013)

The unprecedented expansion of global trade increases the intensity of competition from new market entrants. For example, Panasonic, Samsung, and Sharp must battle for retail shelf space and sales with Vizio, Hisense, and other television manufacturers. Also, the cost of global trade is on the rise. As offshore labor costs increase, global sourcing does not guarantee lower overall cost of goods. In both situations, supply chain management expertise and network flexibility are needed to analyze and respond to these issues. At the same time, globalization can present expansion opportunities. Organizations with flexible supply chain networks that can adapt to the requirements of new markets will be well positioned to grow.

Economic crises such as the recent global recession have a tremendous negative impact on consumer demand and production. Weaker organizations that fail to anticipate the changes, adjust capacity, and reduce inventory levels in their supply chains will not survive. Such was the fate of Circuit City Stores, K-B Toys, and other retailers in 2009. To weather these economic downturns with minimal damage, organizations should build adaptive operating models buoyed by flexible supply chain capacity and a variable cost structure. Also, the use of standardized processes and systems will help the organization rapidly scale or shutter operations based on short-notice demand changes (Cudahy, George, Godfrey, & Rollman, 2012).

With information at their fingertips, today's consumers are empowered to make strong demands on the supply chain. They can review product options, compare prices, and check availability in real-time using mobile devices. This leads to increased expectations for greater product variety, customized goods, off-season availability of inventory, and rapid fulfillment at a cost comparable to in-store offerings. To satisfy these consumer expectations, retailers must be able to leverage inventory as a shared resource and use distributed order management technology to fill orders from the optimal node in the supply chain. Such responsive omnichannel supply chain capabilities separate the retail winners from the losers (Baird & Kilcourse, 2011).

In addition, shrinking product life cycles, the emergence of new technologies to facilitate supply chain transformation, and increases in government regulation of supply chain processes like transportation are compelling reasons to remain nimble. A flexible and responsive supply chain will adapt to these changes with negligible disruption.

Goal 4: Build Network Resiliency

Beyond the business challenges that emerge over time, organizations may also encounter sudden and severe supply chain disruptions. These atypical events—natural disasters, cataclysmic weather, labor strikes, supplier failures, and so on—negatively affect the flow of goods and make the organization vulnerable to financial, reputational, and relational damages. One study estimates that supply chain glitches are associated with an abnormal decrease in shareholder value of more than 10 percent (Hendricks & Singhal, 2003).

Given the cost of disruptions, it is imperative for organizations to manage these supply chain risks. Common predisruption steps include risk identification, risk assessment, and risk reduction. To reduce vulnerability to disruption risks, Sheffi (2005) recommends that organizations collaborate on security and safety issues, build redundancies into their supply chains, and invest in people through cross-training.

In addition to preventative risk management steps, it is imperative to establish disruption management capabilities. Organizations must develop the capabilities to recognize disruptions, overcome them, and redesign processes to reduce future risk (Blackhurst, Craighead, Elkins, & Handfield, 2005). For known risks, it is important to design resilient supply chains that are flexible enough to bounce back quickly from major incidents (Sheffi, 2005). For risks that are unlikely to occur but are potentially catastrophic, supply chain managers must engage in contingency planning and test the plans.

Well known for its configure-to-order (CTO) computer systems, Dell Inc. has structured its supply chain to mitigate risk and recover rapidly from disruptions. The CTO process allows Dell to overcome component shortages by configuring systems in different ways and by enticing customers to specify configurations with components that are readily available. Dell also builds strong long-term relationships with primary suppliers to ensure its priority customer status in times of supply uncertainty. Finally, Dell preemptively qualifies and reviews secondary suppliers to reduce the risk of inventory shortages. Strategies like these minimized the impact of the 2011 Tohoku, Japan earthquake on Dell (de Souza, Goh, Kumar, & Chong, 2011).

Goal 5: Facilitate Financial Success

One of the most important roles of supply chain management is to contribute to the financial success of the organization. Traditional initiatives focus on cost efficiency—streamline stock levels to reduce inventory carrying cost, automate fulfillment operations to minimize labor expense, consolidate orders to cut freight spend, and so on. In contrast, leading organizations use the supply chain to enhance differentiation, increase sales, and penetrate new markets. Their goal is to drive competitive advantage and shareholder value (Anderson, Copacino, Lee, & Starr, 2003).

A dual focus on cost control and revenue generation helps C-level executives recognize the organizational value of supply chain management. As they place more strategic emphasis on supply chain management, capabilities must morph from a series of day-to-day functions to a strategic process with supply chain managers who skillfully manage cross-functional and cross-company complexity. They must understand the connections and interdependencies across the organization and conquer the challenges of managing supplier and customer interfaces (Dittmann, 2012).

Further details regarding supply chain management's role in driving financial success are discussed in the value proposition section.

Evaluation of Supply Chain Management

Supply chain management does not have a long history relative to other business disciplines such as accounting or economics. The term *supply chain management* was first introduced by Keith Oliver of Booz Allen Hamilton in 1982, but did not gain significant traction until the turn of the 21st century (Heckmann, Dermot, & Engel, 2003). However, concepts that underpin supply chain management have been in existence for many decades. For example, today's supply chain strategies continue to draw upon the customer focus of early 20th century catalog retailers and the military's logistics goal of "getting the right people and the appropriate supplies to the right place at the right time and in the proper condition" (U.S. Department of the Army, 1949).

From a business perspective, the origins of supply chain management lie in a wide variety of related but initially fragmented activities. As Figure 1-3 indicates, purchasing, inventory management, warehousing, order processing, transportation, and related functions were conducted independently. Each one had its own budget, processes, priorities, and key performance indicators, but this disaggregated approach was suboptimal and did not lead to lowest total costs.

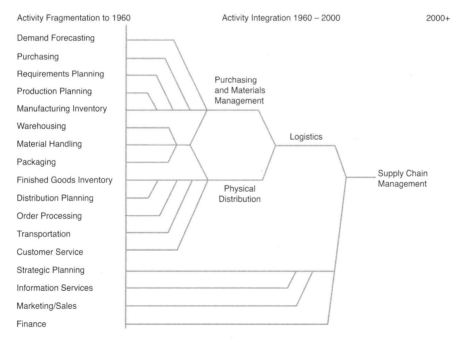

Figure 1-3 The genesis of supply chain management

Eventually, company leaders came to realize the problems of fragmentation and began to integrate related activities. Inbound transportation, purchasing, and production related activities were coordinated in support of manufacturing. Inventory management, order processing, outbound transportation, and related activities comprised the physical distribution function.

Later, these two areas evolved into the logistics function or process that coordinates and integrates the inbound and outbound flows of the organization.

A true supply chain emerges when multiple organizations synchronize their respective processes and adopt a more holistic supply chain management philosophy that includes strategic consideration of related areas. This includes finance, marketing, planning, and technology.

Although the field of supply chain management has rapidly evolved over the last 30 years, many organizations are in the early stages of their supply chain development, and few have fully achieved their desired state of supply chain maturity. This developmental journey is highlighted in Figure 1-4.

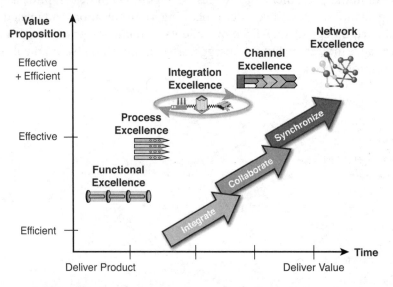

Figure 1-4　The journey to supply chain management

Late adopters of supply chain management must deliberately replace functional silos and cost goals with aligned internal processes. This is often the most challenging aspect of evolving to supply chain management. LaLonde (1999) noted: *"The obstacles to supply chain integration encountered within the organization are far more difficult to overcome than the external challenges."*

After an organization integrates its internal processes and adopts unified cost and service performance targets, focus shifts toward building external relationships and extending the enterprise. Collaboration with key suppliers and customers, robust capabilities, and advanced technologies help the organization drive cross-channel value.

The final step in the maturation process is the development of true network capabilities. A truly dynamic supply chain is needed to support the organizational responsiveness and network resiliency goals discussed earlier. Table 1-2 summarizes the strategic fit and executional capabilities of an organization at each stage of its supply chain development.

Table 1-2 Evolutionary Capabilities

		Functional Excellence	Integrated SCM	Extended Enterprise	Dynamic SCM
		← 1980s →	← 1990s →	← 2000s →	Onward→
Fit with business strategy	Role of supply chain	Meet internal commitments	Meet a customer commitment	Design and fulfill	Design, fulfill, and drive profit
	Extent of influence	Departmental boundaries	Company boundaries	Selected partners	"Ecosystem"/ networks
	Financial focus	Cost	Cost and service	Drive value	Dynamically optimize trade-offs
	Operational focus	Compliance	Interdependence	Collaboration	Agility
	Order management philosophy	First come, first served	Available to promise	Capable to promise	Profitable to promise
	Partner integration	Arm's length	Tight integration	Rationalization (less is more)	Interchangeable
Ability to execute	Supply/demand balancing approach	Produce to a schedule	Fulfill aggregate demand	Forecasting and differentiated fulfillment	Sense, shape, and respond
	Decisioning	Siloed	Team-based	Rapidly address the urgent	Rapidly address the important
	Risk factoring	Afterthought	Buffers in the system	Contingencies and redundancies	Predictive and responsive
	Event horizons	Months	Weeks	Days	Near real-time
	Technology	Standalone applications	MRP/DRP	ERP and bolt-ons ("can plan")	Adaptive layer

		Functional Excellence	Integrated SCM	Extended Enterprise	Dynamic SCM
		← 1980s →	← 1990s →	← 2000s →	Onward→
	Talent	Job functional specialists	Multitasking: Expert in several areas	Career: SCM as a broad profession	Leadership: SCM as a business to be run

Source: Cudahy, G. C., George, M. O., Godfrey, G. R., & Rollman, M. J. 2012. Preparing for the unpredictable. *Outlook: The Online Journal of High-Performance Business*. Retrieved August 8, 2013, from http://www.accenture.com/us-en/outlook/Pages/outlook-journal-2012-preparing-for-unpredictable.aspx.

Value Proposition

The promise of supply chain management is highlighted throughout this introductory chapter. Collectively, the discussion provided in the definition, goals, and evolution sections indicate that supply chain management provides utility to customers, achieves fulfillment goals, and generates shareholder value. Each of these value propositions is discussed in the following sections.

Customer Utility

A review of marketing principles indicates that there are four utilities provided by a business. These five types of usefulness or benefits include form utility, possession utility, place utility, time utility, and quantity utility (Coyle et al., 2013). Form utility—changing the physical characteristics and value of components and parts by assembling them into useful finished goods—is the focus of manufacturing processes. Possession utility is the responsibility of the marketing process. It focuses on facilitating the sale and transfer of ownership of the goods. Generating place, time, and quantity utilities for goods is the mission of supply chain management.

Supply chains generate place utility by moving goods from production points to market locations where demand exists. By having goods readily available in locations that are accessible to interested customers, economic value is added to the goods. For example, moving flu vaccines from the factory to pharmacies and physician's offices in regions where the risk of illness is high, patients get vaccinated and avoid contracting the virus.

Time utility is created by having products available when customers demand them. Supply chain managers must coordinate the movement of inventory from production and storage locations to demand locations as needed. Just-in-time deliveries of transmissions, engines, and other key components are essential for producing cars as scheduled. Similarly, backpacks, pencil sharpeners, and markers must be on-shelf when retailers launch back-to-school advertisements.

Quantity utility ensures that the right amount of product is available to satisfy demand. Supply chain managers must use a combination of forecasting, scheduling, and inventory to achieve quantity utility. Having too much stock increases cost; having too little results in stockouts. When Ford schedules the assembly of 500 Mustang convertibles, Goodyear must deliver 2,000 tires to the factory to support the production schedule.

Place, time, and quantity utilities work hand-in-hand to create value for customers. A great example is the capability of Disney World to temporarily set up refreshment carts along a parade route on a hot day. These dynamically supply chains provide an adequate supply of cold beverages where and when demand exists. Thirsty customers are satisfied, and Disney generates additional sales.

In addition, the issues of product variety, condition, and price are also required to achieve the supply chain value proposition. Thus, to satisfy and retain customers, supply chains must deliver upon the Seven Rights of Fulfillment: *providing the right product, to the right customer, at the right time, at the right place, in the right condition, in the right quantity, at the right cost.*

Fulfillment Success

Achieving the Seven Rights of Fulfillment is possible only if an organization establishes the supply chain capabilities to serve demand better, faster, and cheaper than its competition. Not only is it imperative to focus on effective satisfaction of customer requirements but it is also critical to fulfill demand as efficiently as possible. That is, the organization must minimize supply chain costs subject to its customer service policy to ensure that all parties derive value from the transaction.

Being better than the competition requires an organization to understand customer requirements and develop the supply chain capabilities to support them. Customers purchase goods on the basis of price, quality, delivery, and value-added services. Supply chain management facilitates these purchases by fulfilling demand at optimal performance levels. Thus, an organization must have the right product available in the supply chain with the capability to deliver the goods on time and in full. For example, the Amazon Prime program works only because the company understands demand and positions needed inventory at locations within a 2-day service area of consumers in the program.

Being faster than the competition depends upon the capability to quickly fill and deliver orders. Speed to market is a competitive differentiator for organizations that consistently meet the desired delivery windows of customers. Supply chain managers must also establish the capability and capacity to adjust that speed depending on the situation. A flexible supply chain that supports both premium service requirements (next day or second day fulfillment and delivery) and standard service requirements supports customer needs and creates opportunities for additional business.

Being cheaper than the competition depends on an organization's capability to generate operational efficiencies. Improvement of day-to-day processes through redesign for greater productivity, better asset utilization, and reduction of waste are needed to achieve efficiency. Leveraging the existing resources and expertise of logistics service providers and other capable supply chain partners can also drive efficiency. The imperative is to generate a lower landed cost and lower total cost of ownership than customers receive from the competition.

The capability to concurrently accomplish these better, faster, and cheaper fulfillment goals is not an easy proposition. Conventional wisdom holds that a supply chain can readily provide two of the three desired outcomes. For example, a supply chain can be designed to provide 100 percent in-stock availability and next day delivery, but the cost of achieving this level of service could be crippling. However, leading manufacturers and retailers are working diligently to reach their fulfillment goals in all three areas.

Fulfillment success across all three better, faster, and cheaper goals requires that an organization improve internal processes and strengthen its cross-chain links. Internally, the organization must eliminate unnecessary steps and touch points in the fulfillment process to rationalize product flows. The supply chain managers must also develop collaborative, trusting relationships with suppliers and customers to improve communication, inventory visibility, product flows, and capacity utilization. Both internal and cross-chain cost control initiatives are needed to eliminate excess inventory and waste in the supply chain.

The value proposition and potential payoff for achieving the better, faster, cheaper trifecta of fulfillment goals is significant. The organization with the strongest supply chain in its industry will build a sustainable competitive advantage in the marketplace and generate a higher return on its supply chain investment. These are outcomes worth pursuing for the organization, its supply chain partners, and its shareholders.

Shareholder Value

From a value creation perspective, many organizations limit their financial focus to efficiency in the form of tighter cost control. Given the amount of money spent on logistics and order fulfillment in the supply chain, this is an important goal. However, a pure cost reduction focus misses multiple opportunities to positively affect the income statement. Over the last decade, multiple studies have shown that well-executed supply chain strategies can enhance revenues, improve fixed-capital efficiency, control working capital, and limit tax burdens.

Given the links between supply chain decisions and organizational financial performance of the organization, it is imperative that supply chain managers understand how their actions and resource utilization affect financial statements, profitability, and shareholder value (Wisner, 2011). A concise way to evaluate financial performance is via the Strategic Profit Model highlighted in Figure 1-5.

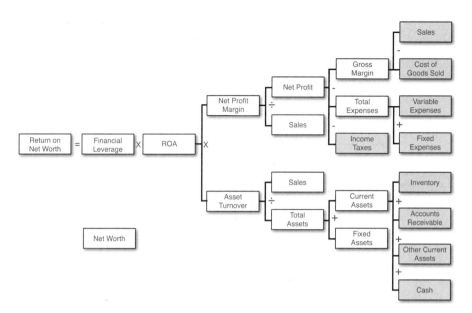

Figure 1-5   The Strategic Profit Model

The model shows how return on net worth is a function of three controllable factors: net profit, asset turnover, and financial leverage. Return on Net Worth is a measure of a company's profitability that reveals how much profit a company generates with the money that the equity shareholders have invested. Of the items identified in the Strategic Profit Model, supply chain strategies typically focus on increasing sales, reducing cost of goods sold, decreasing variable expenses, reducing inventories, and reducing accounts receivables (Stapleton, Hanna, Yagla, Johnson, & Markussen, 2002).

Figure 1.6 takes the general model a step further, linking logical supply chain strategies to their effects on the financial statements of the organization. For example, if a major retailer such as H&M successfully initiates an inventory rationalization strategy, positive outcomes can be achieved. As long as revenues do not decline, the reduced inventory levels in the H&M system will result in lower inventory carrying costs and a positive impact on profits. Likewise, the reduced inventory levels will produce higher inventory turns and produce lower working capital investment. These higher profits and lower investments generate higher return on net worth.

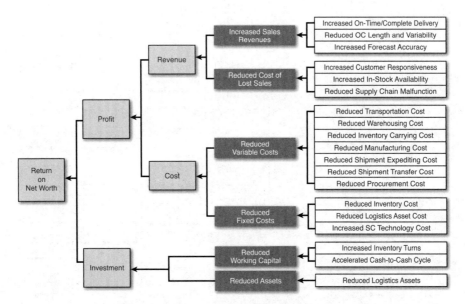

Figure 1-6   Supply chain strategy—Financial performance linkage

To drive shareholder value, supply chain managers must think in terms of Figure 1-6. It is essential to connect the dots between their decisions, the immediate supply chain benefits, and the ultimate effect on organizational financial goals. Doing so will ensure that the supply chain team is making decisions that contribute to the success of the organization and its shareholders.

Structure of the Book

Supply chain management is a complex and increasingly important contributor to organizational success. The foundation descriptions, objectives, and participants discussed in this chapter are essential, but they provide only part of the story. To fully appreciate the opportunities presented via supply chain management, you must understand the plan-buy-make-move-return processes, strategies, technologies, global considerations, and performance tools. The balance of the book covers these essential topics.

Chapter 2: Supply Chain Structure, Processes, and Trade-Offs

The supply chain is made up of essential processes that facilitate the fulfillment of customer demand. Chapter 2 discusses these processes in the context of multidirectional flows of product, information, and money through these supply chain processes, as well as the channels through which demand is served. Each of the plan-buy-make-move-return processes is explained in detail, along with the supply chain trade-off decisions that managers make among these processes.

Chapter 3: Key Strategic Principles

Successful supply chain management depends on integrated planning and development of strategies that promote efficient, effective fulfillment. Chapter 3 begins with an analysis of the guiding principles that supply chain managers must embrace. Next, the strategies for improving supply chain functionality within the organization and coordination across organizations are discussed. The chapter concludes with coverage of barriers to supply chain management success—those potential challenges and pitfalls that can negatively affect performance.

Chapter 4: Supply Chain Information and Technology

Information about customer demand and visibility of inventory are often considered as important as the inventory itself. Chapter 4 addresses the technological capabilities needed to provide cross-chain visibility and support decision making. The chapter begins with a discussion of the role of technology in supply chain planning and execution, including information requirements and capabilities. Also, supply chain software and automatic identification tools are presented. The chapter wraps up with a review of emerging technologies and innovations.

Chapter 5: Managing the Global Chain

The movement of product, information, and funds between countries can be a very challenging proposition. Chapter 5 introduces the global supply chain management activities that support international trade. The chapter covers the challenges, requirements, and external factors that must be considered by supply chain managers. Next, the activities involved in exporting, transporting, and importing goods are discussed. A discussion of key cost considerations wraps up the global supply chain discussion.

Chapter 6: World Class Supply Chain Performance

Supply chain managers must measure performance to evaluate the success of their strategies and operations. Chapter 6 focuses on the use of metrics and related measurement frameworks to ensure compliance with customer requirements and internal goals. Important topics of discussion include the role of measurement, types of measures, trade-off analysis, and measurement systems. The chapter concludes with a discussion of the relationship between supply chain performance and financial outcomes.

Chapter 7: The Supply Chain of the Future

Supply chains are dynamic, evolving entities that must be managed in a forward-thinking fashion. Chapter 7 highlights the emerging issues that will create challenges for supply chain managers and potentially drive change in their processes. Shifting global supply networks, talent shortages, sustainability requirements, omnichannel developments, and rising customer expectations are among the game changers that will shape supply chain strategy in the coming years.

Chapter Summary

A supply chain is more than a simplistic set of links along a linear path from raw material extraction to end consumer use. Instead, this chapter has demonstrated that a supply chain is a very complex network through which materials, information, and money flow between key participants. Highly integrated supply chains dynamically draw upon the capabilities and resources of multiple organizations to ensure the timely and efficient fulfillment of customer requirements.

Integration and synchronization of a supply chain do not occur by happenstance. An organization must make supply chain management a strategic priority, assemble the right talent, build key relationships, and invest in essential technological capabilities. These efforts support the development and coordination of interdependent processes in planning, procurement, conversion, and logistics. All are needed to achieve supply chain excellence.

As this chapter has highlighted, supply chain management is not strictly an in-house function. Even the largest global manufacturers rely on the expertise of external entities to provide essential resources and capacity. These key participants include retailers, distributors, manufacturers, and suppliers who have a financial interest in the products flowing through the supply chain. Also needed are facilitators—logistics services providers, technology companies, indirect materials and equipment suppliers, and so on—who enable the key flows and facilitate the safe, efficient execution of supply chain activities.

The payoffs for investing in supply chain processes and cross-chain relationships are many. As companies move from a functional excellence focus toward integration, collaboration, and synchronization, the supply chain becomes more dynamic and capable of achieving fundamental organizational goals. That is, as its supply chain capabilities mature, the organization can achieve efficient fulfillment, meet customer requirements, respond more effectively to change, become more resilient to disruptions, and improve financial performance. Ultimately, these capabilities translate to greater customer utility, performance versus goals, and shareholder value. These trifold benefits fuel the growing interest in supply chain management among CEOs and corporate boards.

References

Anderson, D. L., Copacino, W. C., Lee, H. L., & Starr, C. E. (2003). *Creating and Sustaining the High-Performance Business: Research and Insights on the Role of Supply Chain Mastery.* Retrieved August 8, 2013, from http://supplychainventure.info/PDF/AccentureSupplyChainMasteryWhitePaper.pdf.

Baird, N., & Kilcourse, B. (2011). *Omni-Channel Fulfillment and the Future of the Retail Supply Chain.* Retrieved August 7, 2013, from http://www.scdigest.com/assets/reps/Omni_Channel_Fulfillment.pdf.

Ballou, R. H., Gilbert, S. M., & Mukerjee, A. (1999) Managing in the new era of multi-enterprise supply chains. *Proceedings of the Twenty-Eighth Annual Transportation and Logistics Educators Conference.* Eds. Bernard J. LaLonde & Terrance L. Pohlen. Toronto: Council of Logistics Management, 23–37.

Blackhurst, J., Craighead, C. W., Elkins, D., & Handfield, R. B. (2005) An empirically derived agenda of critical research issues for managing supply-chain disruptions. *International Journal of Production Research*, 43(9), 4067–4081.

Bradley, P. (2013) Collaboration bears fruit. *CSCMP's Supply Chain Quarterly*, 7(2), 34–36.

Christopher, M. L. (1992) *Logistics and Supply Chain Management.* London: Pitman Publishing.

Cooke, J. A. (2013) Kimberly-Clark connects its supply chain to the store shelf. *DC Velocity*, 11(5): 53–55.

Council of Supply Chain Management Professionals. (2010) *Supply Chain Management Terms and Glossary.* Retrieved August 2, 2013, from http://cscmp.org/sites/default/files/user_uploads/resources/downloads/glossary.pdf.

Council of Supply Chain Management Professionals. (2011) "Supply Chain Management Definition." Retrieved August 2, 2013, from http://www.careersinsupplychain.org/what-is-scm/definition.asp.

Council of Supply Chain Management Professionals. (2013) *24th Annual State of Logistics Report: Is This the New Normal?* Oak Brook, IL: Council of Supply Chain Management Professionals.

Coyle, J. J., Langley, C. J., Novack, R. A., & Gibson, B. J. (2013) *Supply Chain Management: A Logistics Perspective.* Mason, OH: South-Western Cengage Learning.

Cudahy, G. C., George, M. O., Godfrey, G. R., & Rollman, M. J. (2012) Preparing for the unpredictable. *Outlook: The Online Journal of High-Performance Business.* Retrieved August 8, 2013, from http://www.accenture.com/us-en/outlook/Pages/outlook-journal-2012-preparing-for-unpredictable.aspx.

de Souza, R., Goh, M., Kumar, M., & Chong, J. (2011). *Combating Supply Chain Disruptions: Lessons Learned from Japan 2011*. Retrieved August 8, 2013, from http://www.go2uti.com/c/document_library/get_file?uuid=cb104971-5409-4a3e-8ded-de271d19ad2e&groupId=31941.

Dittmann, J. P. (2012) Start with the customer! *CSCMP's Supply Chain Quarterly*. Retrieved August 8, 2013, from http://www.supplychainquarterly.com/topics/Strategy/20121217-start-with-the-customer/.

Gartner. (2013a) *Gartner Announces Rankings of its 2013 Supply Chain Top 25*. Retrieved August 8, 2013, from http://www.gartner.com/newsroom/id/2494115.

Gartner. (2013b) *IT Glossary*. Retrieved August 2, 2013, from http://www.gartner.com/it-glossary/supply-chain-management-scm/.

Gibson, B. J., Mentzer, J. T., & Cook, R. L. (2005) Supply chain management: The pursuit of a consensus definition. *Journal of Business Logistics*, 26(2), 17–25.

Heckmann, P., Dermot, S., & Engel, H. (2003) *Supply Chain Management at 21: The Hard Road to Adulthood*. Retrieved August 8, 2013, from http://www.boozallen.com/media/file/supply-chain-management-at-21.pdf.

Hendricks, K. B., & Singhal, V. R. (2003) The effect of supply chain glitches on shareholder wealth. *Journal of Operations Management 21*(5), 501–522.

Institute for Supply Management. (2010) *Supply Management Defined*. Retrieved August 2, 2013, from http://www.ism.ws/tools/content.cfm?ItemNumber=5558.

LaLonde, B. J. (1997) Supply chain management: Myth or reality? *Supply Chain Management Review*, 1(Spring), 6–7.

LaLonde, B. J. (1999) The quest for supply chain integration? *Supply Chain Management Review*, 2(4), 7–10.

Lummus, R. R., & Vokura, R. J. (1999) Defining supply chain management: A historical perspective and practical guidelines. *Industrial Management & Data Systems*, 99(1), 11–17.

Mentzer, J. T., DeWitt, W., Keebler, J. S., Min, S., Nix, N. W., Smith, C. D., & Zacharia, Z. G. (2001) Defining supply chain management. *Journal of Business Logistics*, 22(2), 1–26.

Nike, Inc. (2013) Global Manufacturing. Retrieved August 9, 2013, from http://manufacturingmap.nikeinc.com/.

Partridge, A. R. (2010) Managing a customer-driven supply chain. *Inbound Logistics*. Retrieved August, 8, 2013, from http://www.inboundlogistics.com/cms/article/managing-a-customer-driven-supply-chain/.

Porter, M. E. (1985) *Competitive Strategy: Techniques for Analyzing Industries and Competitors*. New York: The Free Press.

Ramsey, J. G. (2005) The real meaning of value in trading relationships. *International Journal of Operations & Production Management*, 25(2), 549–565.

Sheffi, Y. (2005) *The Resilient Enterprise*. Cambridge, MA: The MIT Press, 269–285.

Stapleton, D., Hanna, J. B., Yagla, S., Johnson, J., & Markussen, D. (2002) Measuring logistics performance using the strategic profit model. *International Journal of Logistics Management*, 13(1), 89–107.

Stock, J., & Boyer, S. (2009) Developing a consensus definition of supply chain management: A qualitative study. *International Journal of Physical Distribution & Logistics Management*, 39(8), 690–711.

Sweeney, E. (2011) Towards a unified definition of supply chain management. *International Journal of Applied Logistics*, 2(3), 30–48.

U.S. Department of the Army. (1949) *FM 100-5, Field Service Regulations-Operations*. Retrieved August 7, 2013, from http://www.cgsc.edu/CARL/docrepository/FM100_5_1949.pdf.

Wisner, P. (2011) Linking supply chain performance to a firm's financial performance. *Supply Chain Management Review*. Retrieved August 20, 2013, from http://www.scmr.com/article/linking_supply_chain_performance_to_a_firms_financial_performance.

2

SUPPLY CHAIN STRUCTURE, PROCESSES, AND TRADE-OFFS

A true supply chain is a network of organizations that provides seamless flows via effective processes. When these capabilities are not present, the supply chain exists in name only, and the outcome is a makeshift collection of activities that will be susceptible to errors, disruptions, and outright failure. Hence, supply chain managers must work to continuously improve each process and integrate them to ensure that products, information, and payments move efficiently across the network.

The importance of flows and processes are highlighted throughout the opening chapter in relation to supply chain definitions, goals, and value propositions. They are more than abstract concepts; flows and processes are tangible elements of supply chain strategy, capabilities, and performance. To fully understand their contribution to supply chain success, the impact of each flow and the foundational roles of each process are addressed in this chapter.

The relative importance of and synergy between these two supply chain elements can be inferred from simple dictionary definitions. According to the *Merriam-Webster Dictionary* (2013), a flow endeavors "to move in a continuous and smooth way," whereas a process is "a series of actions that produce something or that lead to a particular result." Similarly, the Council of Supply Chain Management Professionals (CSCMP, 2010) defines a process as "a series of time-based activities that are linked to complete a specific output."

The ideal result or output of a supply chain process is a steady flow of product, information, and financial resources that results in a profitable satisfaction of customer demand. This can be accomplished when processes are properly aligned within and across organizations in the supply chain. Hammer (2001) pointed out that within the integration of business processes across firms is where the real "gold" can be found. Lambert (2004) noted that supply chain management (SCM) is, in essence, the integration of key business processes from end users through original suppliers. Thus, a process view is fundamental to SCM.

Two additional considerations must be included in the discussion of supply chain flows and processes. First, the structure of a supply chain and the channels through which product flows must be considered. The structure of the supply chain network determines the channel options through which products flow from sources to end users. Also, trade-offs must be made between processes and stakeholders in the supply chain. How these cross-functional interactions are carried out will influence product flows, process costs, and supply chain competitiveness.

This chapter provides valuable insights into these four important supply chain elements. Supply chain flows will be discussed first because these requirements drive development of the structure, processes, and trade-offs. Next, an important overview of supply chain structures is provided. A detailed discussion of the five primary supply chain processes follows, with emphasis on the purpose and activities of each process. The chapter concludes with an analysis of the ever-present trade-offs that must be made within and across the supply chain processes. Managers must understand the cross-chain effects of their strategic and operational decisions.

Supply Chain Flows

A key objective of SCM is to facilitate smooth, coordinated movement across the supply chain. In general, there are three types of flows in a supply chain: products/services, information, and payments. Figure 2-1 represents a simplified supply chain with a limited number of participants and the three flows occurring downstream and upstream. Understanding these essential flows is critical to effectively managing supply chain processes.

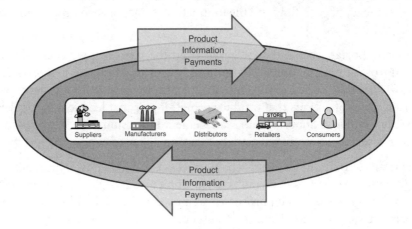

Figure 2-1   Supply chain flows

Product and Related Service Flows

Materials movement has traditionally been the focus of logisticians and is still an important element in SCM. To meet customer needs, the ordered products should be delivered a timely, reliable, and damage-free manner. Transportation and warehousing are critical for achieving this outcome. It must also be realized that products and related packaging might flow backward due to customer returns, unsold products, product repairs, and packaging reuse/recycling. In fact, with the fast growth of e-commerce and enhanced sustainability initiatives, reverse logistics—the logistics function that deals with the reverse flows—is gaining more attention from companies for its strategic and financial importance.

Information Flows

Timely data and knowledge sharing is another essential factor for achieving effective SCM. A key aspect of SCM is the ability to make strategic decisions quickly, based on accurate information. Every supply chain has an information chain that parallels the flow of product (Andel, 1997). Without information relayed at the right time to the right place, there are no purchase orders, no shipment messages, no payments, no coordinated marketing and sales efforts, and the supply chain will shut down (Zuckerman, 1998). Therefore, substantial information exchange among supply chain stakeholders is vital for a supply chain to function efficiently and effectively (Mentzer, 2004).

Both forward and backward information flows are needed. Forward information flows in the supply chain can take many forms such as advance shipment notices, order status information, and inventory availability information. Reverse information flow can carry information about the market demand, customer preferences to upstream supply chain members from distributors/retailers or consumers. Both are essential for a smooth running supply chain.

Monetary Flows

The third flow focuses on financial transactions, particularly the movement of payments. Traditionally, payment flows were viewed as one-directional—upstream—in the form of payment for goods, services, and orders received. However, as the reverse logistics volume increases, managing the associated payment flows from manufacturers or retailers to customers have also been put on supply chain managers' to-do list. Timely refunds in the reverse logistics context can significantly affect customer satisfaction; no customer wants to wait for a long time to get money back.

No matter how big or small, organizations pay special attention to their capital resources. It must be realized that a company's supply chain financial flows have significant impacts on working capital. Therefore, in addition to order cycle, companies measure supply chain performance in terms of their cash-to-cash cycle. This metric, also called the cash conversion cycle, reveals the time it takes for cash to flow back into a company after capital has been spent for raw materials (CSCMP, 2010). Some companies such as Dell successfully achieve a negative cash conversion cycle by collecting money from customers before they have to pay their vendors or suppliers. For example, the period between collection from Dell's customers and payment to its suppliers may be 30 to 45 days. The cash can be used for other purposes such as financial investments or funding for product development.

Supply Chain Structure

A basic supply chain structure, as depicted in Figure 2-1, has a very limited number of participants. However, the idea of direct linear links between raw materials suppliers and consumers oversimplifies the true structure of supply chains. In reality, most supply chains are complex, dynamic, and somewhat unique networks that involve many different parties. As stated in Chapter 1, an organization's supply chain structure is influenced by its industry, geographic scope of activity, supply base, product variety, fulfillment methods, and demand patterns.

Consider, for example, the structure of the LCD television supply chain depicted in Figure 2-2. From the product manufacturer's perspective, the firms whose inputs feed into the assembly operations are positioned upstream. Firms that move the finished televisions through the supply chain channels to consumers are positioned downstream. A supplier that directly provides component parts to the focal company is referred as the first-tier supplier. The first-tier supplier may also source from its own upstream suppliers, which are second-tier suppliers. Hence, the television manufacturer is both a downstream customer and upstream supplier in the supply chain. This dual customer and supplier role is also true for most organizations in a supply chain.

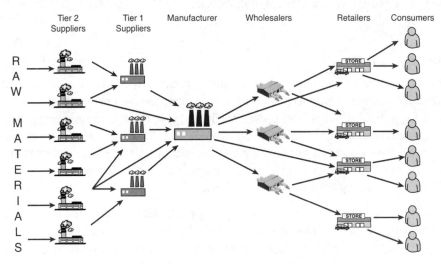

Figure 2-2   Supply chain structure

Figure 2-2 still does not fully capture the complex structural composition of most supply chains. Some upstream organizations are both first-tier and second-tier suppliers to the television manufacturer. Also, the downstream consumers have the capability to purchase televisions from multiple retailers, wholesalers, and direct from manufacturers. In addition, Figure 2-2 does not clearly depict the supply chain participants who provide supporting functions. For example, a third-party financial provider may be providing financing, assuming some risks, and offering financial advice; a third-party logistics service provider (3PL) may be performing the logistics activities to facilitate the movement of the goods; and a market research firm may be providing information and analysis on end consumers (Mentzer, 2001).

For most organizations and industries, a network perspective is appropriate for their supply chain structures. The term "network" implies a complex structure, in which organizations are cross-linked with two-way interchange of products, information, and payments. These supply chain network structures include multiple processes that extend across organizational boundaries. Managers must treat their network structures as systems because all processes within the network structure need to be understood in terms of how they interact with other processes. No organization is isolated in a network structure; its inputs and outputs are affected by the actions of other network participants.

Another distinct characteristic of a supply chain structure from a traditional management philosophy is the relationships among supply chain members. A supply chain perspective shifts the business structure from a loosely linked group of independent businesses to a coordinated multi-enterprise effort focused on supply chain efficiency improvement and increased competitiveness (Bowersox, Closs, & Cooper, 2007). In contrast to traditional adversarial business relationships, a fundamental belief in SCM is that collaborative behavior will reduce risk and greatly improve efficiency of the overall supply chain process. Such collaboration requires mutual trust and commitment from supply chain members, which in turn will facilitate free information sharing and coordinated actions across the network.

Overall, supply chain structures are growing more interdependent and challenging due to growth of global trade, product variety, and e-commerce. For example, many organizations operate in omnichannel environments in which their customers desire the option to purchase and receive products through multiple channels. Hence, the structure of the supply chain must support fulfillment through retail stores, e-commerce sites, wholesalers/distributors, and manufacturers. Supply chain managers must develop flexible multichannel fulfillment structures to ensure the timely and efficient delivery of products, information, and payments across the network.

Supply Chain Processes

The process paradigm is a progressive alternative to a more traditional division of supply chain activities into functional units, divisions, or department configurations (Cooper et al., 1997). The focus of the functional approach was often associated with a myopic view of the activity's engagement with other activities within and across firms. The process approach broadens the focus of the organization, promoting extensive interaction and collaboration to meet customer needs and create value.

Implementing a process management approach within and across supply chain stakeholder organizations makes the transactions and relationship structures more efficient and effective. However, it is important to note that adopting a process focus is not an easy proposition. It can be a complex undertaking because supply chains have both strategic and operational processes. Also, the scope varies from primary processes to subprocesses within the overall supply chain process of fulfilling customer demand. Unless these processes are all properly identified, explicitly understood, and cohesively managed, supply chain success will be fleeting.

The purpose of this section is to explain the five primary processes in a supply chain and discuss the relationships between them. This can be accomplished through a commonly accepted framework based on the Supply Chain Operations Reference (SCOR) Model. The SCOR Model uses a common set of definitions and metrics to enable an organization to understand its supply chain processes and evaluate their performance (Supply Chain Council, 2010). Figure 2-3 indicates how the plan, source, make, deliver, and return processes occur within organizations and create natural linkages between organizations.

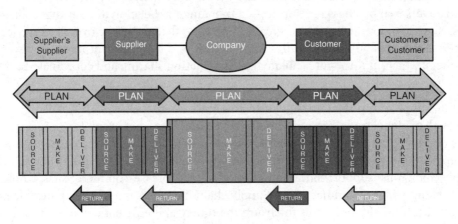

Figure 2-3 Supply chain processes

Plan: Integrating Demand and Supply

Planning involves the creation of an overall strategy for synchronizing supply with demand in supply chains. This involves developing plans for managing resources, flows, and relationships for optimal cost and service across the supply chain (Gibson, 2005). According to the Supply Chain Council (2010), this process includes gathering customer requirements, collecting information on available resources, and balancing requirements and resources to determine planned capabilities and resource gaps. This is followed by identifying the actions required to correct any gaps.

The Role of Demand Planning

The importance of effective demand management cannot be overemphasized because all decisions made in other business processes are based on the outcomes generated from the demand planning process. In essence, demand is the ultimate factor that drives all supply chains. Mentzer (2004) suggested that if we take a supply chain perspective, the supply chain has only one point of independent demand, which is the amount of product demanded, by time and location, by the end-use customer of the supply chain. Whether this end-use customer is a consumer shopping in a retail store or online, or a business buying products for consumption in its own operations, these end-use customers determine the true demand for the product that will flow through the supply chain.

The company in a supply chain that serves the end-use customer experiences this independent demand. All other companies in the supply chain experience a demand that is tempered by the fulfillment and purchasing policies of other companies in the supply chain. Such demand is referred to as derived demand or dependent demand because it is derived from what other companies in the supply chain do to meet their demand from their immediate customer (Mentzer, 2004). Therefore, only one echelon in any given supply chain is directly affected by independent demand, and the rest are affected by derived demand. It is, thus, important to take an integrative perspective when performing supply chain planning. A shared demand planning process coordinates and orchestrates the flow movements seamlessly from one process to the next (Harrison & van Hoek, 2011).

Because demand is the trigger of all supply chain activities, demand planning directly affects other supply chain processes. Procurement needs accurate demand information to obtain the right raw materials in the right quantity at the right time, manufacturing needs correct demand information for efficient production scheduling, and the logistics process requires accurate demand information to maintain appropriate inventory levels and arrange timely deliveries.

Key Elements of the Demand Management Process

Demand management is the function of recognizing all demands for goods and services to support the marketplace. It consists of several major components.

Forecasting

Customer demand is rarely perfectly stable, so businesses must forecast demand to properly position inventory and other resources. Forecasting is an important element of demand planning because it provides an estimate of future demand and the basis for planning and sound business decisions. Accurate demand forecasts allow the purchasing department to order the right amount of products, the operations department to produce the right amount of products, and the logistics department to deliver the right amount of products. In contrast, inaccurate forecasts will lead to imbalance between supply and demand, and cause consequences such as stockouts, lost sales, high costs of excess inventory and obsolescence, material shortages, poor market responsiveness, and poor profitability. For example, when Nintendo first introduced its Wii gaming console, its sales greatly exceeded expectations and the company struggled to meet the unexpectedly high demand for the product.

Because all companies face demand uncertainty and customer volatility, some discrepancy between a forecast and actual demand is to be expected. Therefore, the goal of good forecasting is to minimize the deviations between the forecast and actual demand. To generate an accurate forecast, factors that influence demand must be taken into consideration. Also, buyers and sellers should share relevant information on supply and demand to facilitate accurate decision making.

Supply chain managers must realize that high-quality forecasting benefits not only the focal company but also its supply chain partners. The benefits of better forecasts are lower inventories, reduced stockouts, smoother production plans, reduced costs, and improved customer service. These benefits are achieved by reducing the bullwhip effect, which refers to a trend of larger and larger swings in upstream inventory in response to relatively small changes in demand for a product when moves back in the supply chain because each supply chain participant makes its own forecasts and builds an inventory buffer called safety stock.

It has been suggested that forecasting is the combination of art and science because no technique can provide 100 percent forecasting accuracy, but experience often helps improve forecast accuracy. Generally speaking, forecasting methods fall into two main categories: qualitative and quantitative. Both methods rely on past demand data, so these forecasting methods will generate less-accurate results as forecasts are projected further into the future. Therefore, it is often suggested to use the combination of both qualitative and quantitative forecasting methods.

Qualitative forecasting methods use judgmental evaluations and intuition as the basis and are often used when current data are unavailable or limited (for example, new product information). These methods have relatively low cost, but their efficiency largely depends on the experience and skills of the forecaster and the amount of information available. The inputs of qualitative forecasting can come from different entities such as internal and external executives, salespeople, and customers.

Quantitative forecasting methods utilize mathematical models that are based on historical data and can include causal variables to forecast demand. Time-series forecasting and cause-and-effect forecasting are the two major approaches for quantitative forecasting methods. Time series forecasting is based on the assumption that the future is an extension of the past; therefore, historical data can be used to predict future demand. Cause-and-effect forecasting assumes that one or more factors can affect demand and thus can be used to predict future demand.

Although all forecasting techniques seek to accurately forecast demand, the common thread is that they will ultimately be wrong. The key to successful forecasting is to minimize the error between actual demand and forecasted demand. Although this sounds simple, many factors can arise in the marketplace that will influence demand contrary to the forecast. However, forecasts are necessary because they serve as a plan for both marketing and operations to set goals and develop execution strategies. These goals and strategies are developed through the sales and operations planning (S&OP) process.

Marketing and Sales Planning

An effective demand management process helps an organization develop more feasible marketing and sales plans. Historically, organizations developed several functional forecasts for the same products for the same time period. It was quite common for a manufacturer to have a financial forecast, a manufacturing forecast, a marketing forecast, and a distribution forecast. In most cases, these functional forecasts did not agree, which could cause internal conflicts in communication and execution. Obviously, it is necessary for a company to arrive at a forecast internally that all functional areas agree upon and can execute.

To achieve this unified plan, organizations are embracing S&OP processes. As shown in Figure 2-4, the S&OP process usually consists of five steps:

1. A statistical forecast of future sales is developed using one or more forecasting techniques.

2. In the demand planning activity, the sales and/or marketing department(s) review the forecast and make necessary adjustments based on promotions of existing products, the introduction of new products, or the elimination of products.

3. In the supply planning phase, operations departments (manufacturing, warehousing, and transportation) analyze the sales forecast to determine whether existing capacity is adequate to handle the forecasted volumes in terms of both quantity and timing.

4. A pre-S&OP meeting takes place, involving individuals from sales, marketing, operations, and finance. This meeting will review the initial forecast and any capacity issues emerged during previous steps. Initial attempts will be made during this meeting to solve capacity issues by attempting to balance supply and demand.

5. An executive S&OP meeting will take place to finalize the decisions regarding sales forecasts and capacity. This is where the top executives from the various functional areas agree to the forecast and convert it into the operating plan for the company. Consensus among the various functional areas is critical in this meeting. Decisions regarding trade-offs between revenue and costs are made in this meeting. Once the final plan is approved, it is important that the appropriate metrics are put in place for each functional area to encourage compliance.

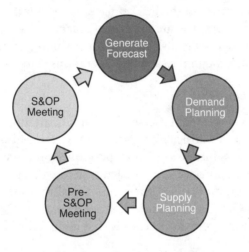

Figure 2-4   The sales and operations planning process

This S&OP process brings all relevant parties onto the same page, promoting coordinated behavior in developing and executing marketing and sales planning. The resulting decisions and actions will be more accurate than the traditional multiplan approach.

Understanding Customer Requirements

One objective of demand management is to enhance the ability of firms throughout the supply chain to collaborate on activities related to the flow of product, services, information, and finances. The desired end result should be to create greater value for the end user or consumer. Because demand management utilizes rich information about customers, it is uniquely positioned to communicate with supply chain members about customer requirements. The following actions help supply chain stakeholders collectively focus on satisfying customers and solving their problems:

- Gathering and analyzing knowledge about consumers, their problems, and their unmet needs
- Sharing with other supply chain members knowledge about consumers and customers, available technology, and logistics challenges and opportunities
- Identifying partners to perform the functions needed in meeting customer needs
- Assigning necessary functions to the supply chain member who can perform them most effectively and efficiently
- Developing products and services that solve customers' problems
- Developing and executing the best logistics, transportation, and distribution methods to deliver products and services to consumers in the desired format

Interfacing Between the Marketplace and the Manufacturing Planning and Control System

Despite being far removed from the marketplace, the manufacturer often makes the critical determinations of what, where, when, and how many units to produce for sale in the marketplace. The problem is magnified when other supply chain participants distort the real marketplace demand with their forecasts and actions. Take a new product as an example. When it is first launched, the end-user demand is at its peak, and opportunities for profit margins are greatest. However, the manufacturer often does not have sufficient capacity to meet the demand, thus creating true product shortages. During the same time, distributors and resellers tend to over-order, often creating substantial phantom demand. As production begins to increase, the manufacturer ships product against this inflated order situation and books sales at the premium, high-level launch price. As channel inventories begin to grow, price competition sets in, as do returns from retailers due to product overages. This further depresses demand for the product, and the manufacturer will suffer from product oversupply.

Such situations are not uncommon and are largely due to the industry's planning processes and systems, which are primarily designed to use previous period demand as a gauge. Because much of the previous period's demand was represented by the previously mentioned phantom demand, forecasts are distorted. Because a large portion of the products are sold during the declining period of profit opportunity, the value creation opportunities for supply chain members suffer significantly. Therefore, understanding and managing market demand are essential to business success, and demand management is the key activity that addresses the possible disconnect between manufacturing and demand at the point of consumption.

Connecting Demand Management and Supply Chain Activities

The S&OP process helps companies restructure their planning processes to arrive at a consensus forecast internally. It is also beneficial for members of a supply chain to agree upon a consensus forecast. More important, a coordinated demand management process across supply chain members can foster supply chain integration. Collaborative Planning, Forecasting, and Replenishment (CPFR) is such a practice that brings various forecasts and plans from different supply chain members into synchronization. Using this approach, retailers, distributors, and manufacturers can utilize available Internet-based technologies to collaborate on operational planning through execution.

Simply put, CPFR allows trading partners to agree to a single forecast for an item where each partner translates this forecast into a single execution plan. This replaces the traditional method of forecasting, in which where each trading partner developed its own different forecast for the same item. The first attempt at CPFR was between Wal-Mart and Warner-Lambert (now a part of Johnson & Johnson) in 1995 for its Listerine product line. In addition to rationalizing inventories of specific line items and addressing out-of-stock occurrences, these two companies collaborated to increase their forecast accuracy, so as to have the right amount of inventory where it was needed, when it was needed. The Voluntary Inter-industry Commerce Standards Committee (VICS) became involved in 1998 and is a major advocate for this initiative.

Figure 2-5 shows the CPFR model as a sequence of several business processes that involve the consumer, retailer, and manufacturer. The four major processes are strategy and planning, demand and supply management, execution, and analysis. Two aspects of this model are important to note. First, it includes the cooperation and exchange of data among business partners. Second, it is a continuous, closed-loop process that uses feedback (analysis) as input for strategy and planning. CPFR emphasizes a sharing of consumer purchase data or point-of-sale data as well as forecasts among and between trading partners for the purpose of helping to manage supply chain activities. From these data, the manufacturer analyzes its ability to meet the forecasted demand. If it cannot meet the demand, a collaborative effort is undertaken between the retailer and manufacturer to arrive at a mutually-agreed-upon forecast from which execution plans are developed. The strength of CPFR is that it provides a single forecast from which trading partners can develop manufacturing strategies, replenishment strategies, and merchandising strategies.

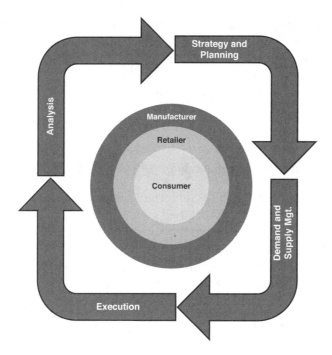

Figure 2-5   Collaborative Planning, Forecasting, and Replenishment (CPFR)

Source: Procuring Goods and Services

Sourcing is the purchase process related to effective supply management, and it is also referred as procurement. CSCMP (2010) defines procurement as the activities associated with acquiring products or services. In practice, managers often use procurement, purchasing, and sourcing interchangeably. Procurement involves supplier selection, transaction term or contract development, procurement of key inputs, delivery receipt and verification, payment and performance control (Gibson, 2005).

The Role of Procurement in SCM

Historically, purchasing was perceived as a clerical or low-level managerial activity responsible for receiving requisitions and issuing purchase orders. The role of purchasing was to obtain the desired resource at the lowest possible purchase price from a supplier. However, this traditional view of purchasing has changed substantially in the past several decades. Today, the primary goals of procurement are to ensure uninterrupted flows of raw materials and parts at the lowest total cost, to improve quality of the finished goods produced and to maximize customer satisfaction. A holistic and comprehensive acquisition strategy is required to meet the organization's strategic objectives.

When elevated to a strategic level, procurement is drastically different from the traditional purchasing in different ways with emphasis on the following:

- Consolidation and leveraging of purchasing power
- Value rather than acquisition cost alone
- More meaningful supplier relationships
- Process improvement
- Enhanced teamwork and collaboration with supply chain partners

At the minimum level, procurement is important to most companies simply because of its sheer volume. For most firms, purchased goods and services are among the largest cost elements. According to the Annual Survey of Manufactures, conducted by the U.S. Census Bureau, manufacturers spent more than 50 percent of each sales dollar on raw materials from 1977 to 2006 (Wisner, Tan, & Leong, 2011). Purchase of raw materials actually exceeded the value added through manufacturing. Therefore, a tremendous opportunity exists for potential cost-savings related to procurement.

More important, procurement's strategic importance in SCM is reflected in its potential to provide competitive advantage. The procurement process can give organizations a sustainable advantage through lower costs, higher quality, or better service for internal or external customers. A strong relationship with trading partners and the implementation of a supply management process that integrates procurement with other business functions is essential for increasing the efficiency and effectiveness of the supply chain.

Key Elements of the Procurement Process

Today's procurement is much different from the traditional purchasing function. It involves a broader range of activities and often has strategic implications for the organization. These sequentially implemented elements include the following.

Develop Procurement Objectives and Organization

The products and services purchased by a company are not all the same. Therefore, the attention and strategies used should also be differentiated for these different types of items. Based on the value and risk associated with the purchased items, purchased items can be categorized into four types:

- Generics are low-risk, low-value items and services that typically do not enter the final product. Items such as office supplies and maintenance, repair, and operating (MRO) items are examples of generics. Procurement focus for these products is to streamline the procurement process to reduce the cost.

- Commodities are items or services that are low in risk but high in value. Basic production materials, packaging, and transportation services are examples of commodities that enhance the profitability of the company but pose a low risk because they are readily available from many sources in the market. Commodities are not unique, so there is little brand distinction, and price is a significant distinguishing factor. Also, because freight and inventory are major procurement cost considerations for commodities, the procurement strategies used for commodities include volume purchasing to reduce price and just-in-time systems to lower inventory costs.

- Distinctives are high-risk, low-value items and services such as engineered items, parts that are available from only a limited number of suppliers, or items that have a long lead time. These products pose a threat to continued operation and/or have a high procurement cost. The strategic focus for distinctives is developing a standardization program to eliminate or reduce the uniqueness of the distinctives, thereby changing these items to generics.

- Criticals are high-risk, high-value items that give the final product a competitive advantage in the marketplace. Criticals can determine the customer's ultimate cost of using the finished product. The procurement focus for criticals is to strengthen their value through use of new technologies, simplification, close supplier relations, and/or value-added alterations. Procurement innovation can help the critical item provide greater market value to the finished product. Given their importance, supply chain managers should direct greater resources and attention toward procurement of criticals rather than generics.

Buying situations are grouped into three types. The first is the procurement of capital goods that may represent a long-term investment for an organization and may require significant financial planning. The second is rebuys, with standard rebuys being repeat purchases that are identical to historical purchases and modified rebuys that involve some order variation. The third is the procurement of MRO items that are needed for the continuing operation of the company and its supply chain activities. The involvement level and inputs from procurement professionals such as energy, time, and information search vary among these buying situations.

Supplier Selection and Negotiation

Although companies use different criteria to select suppliers, a few factors are commonly used by procurement professionals to evaluate potential suppliers: quality, price, reliability, and risk, capability, and financial considerations.

Quality is usually listed as the most important supplier selection criterion. As was indicated earlier, quality often refers to the specifications that a user desires in an item such as technical specifications, chemical or physical properties, or design. Quality refers to the degree to which the purchased item meets the user's requirements on a set of defined characteristics. Quality may also include additional factors such as life of the product, ease of repair, maintenance requirements, ease of use, and dependability.

The price of purchased items will directly affect the buying firm's cost structure and profitability. Therefore, price is no doubt an important factor to consider when selecting a supplier. However, instead of solely focusing on the purchase price, companies are now emphasizing the lowest Total Cost of Ownership (TCO). TCO includes all the direct and indirect costs associated with an asset or acquisition over the entire life cycle of the product or service. It includes not just the purchase price but also such things as transportation, handling and storage, damage and shrinkage, taxes, insurance, and redistribution costs. It may also include installation, upgrade costs, training, support, service, maintenance, downtime, retirement costs and/or disposal (CSCMP, 2010).

Reliability means on-time delivery and consistent product performance history. Consistent, on-time deliveries are key to preventing production disruptions resulting from longer-than-expected lead times. The performance of the procured product directly affects the quality of the final product and its market performance. The supplier's warranty and claim procedure is also often considered as a reliability measure.

All supply chain professionals want to reduce uncertainties. Supply uncertainties and variations in delivery lead times and prices can all pose threats to the buying firm's operations or incur additional costs. Therefore, risk associated with a potential supplier is often another important selection criterion.

To compete in the dynamic market environment, the buying firms often highly value potential suppliers' capabilities such as their production facilities and capacity, technical capability, management and organizational capabilities, and operating controls. These factors indicate the supplier's capability to provide a needed quality and quantity of material in a timely manner. The evaluation includes not only the supplier's physical capability to provide the material the user needs but also the supplier's capability to do so consistently over an extended time period.

In addition to the purchase price, the buying firm often evaluates the supplier's financial position. Financially unstable suppliers create the risk of possible disruptions to the ongoing supply of material. With the trend toward consolidating the number of suppliers, the financial failure of a supplier can become a major problem and source of disruption in a supply chain.

Besides the previously mentioned factors, companies also use other criteria in the supplier selection process, such as the supplier's geographical location, the supplier's attitude, the impression or image that the supplier projects, training aids, and technology support. The relative importance of these selection factors depend upon the specific procurement situation.

Negotiation is particularly effective when the buyer is interested in a strategic alliance or long-term relationship. The negotiation process can be time-consuming, but the potential benefits can be significant in terms of price and quality. More important, negotiation in fact is an effective communication approach to better understanding each other's needs and requirements, which is fundamental to developing long-lasting supply chain partnership.

Execution and Control

Although all firms are different and have different requirements for the procurement items, a basic four-step approach can be used and adapted to a company's particular needs:

1. **Determine the type of purchase**—Identifying the type of purchase will often determine the complexity of the entire process. The straight rebuy situation would involve only minimal additional effort while the modified rebuy and new buy situations will involve significantly more effort.

2. **Determine the necessary levels of investment in terms of time and information**—Time is expended by the individuals involved in making the purchase, and information can be both internal and external to the firm. Generally speaking, the more complex and important the purchase, the more time and information are needed for the procurement process to be effective.

3. **Perform the procurement process**—This includes performing those activities necessary to effectively make a purchase and satisfy the user's requirements. Depending on the buying situation, the actual procurement process can be complex. Measuring the actual investment and how well a user's needs are satisfied is important to the future evaluations of the purchase.

4. **Evaluate the effectiveness of the sourcing process**—This is a control step that examines whether the user's needs are satisfied and whether the investment is necessary. If it is found that the sourcing process is less effective than expected, the causes should be identified and corrective actions taken.

Although the procurement process is complex, it can be managed effectively as long as the managers develop a systematic approach for implementing it. A key factor in achieving efficiency and effectiveness in this area is the development of successful supplier relationships. In fact, many professional procurement managers agree that strong supplier relationships are needed to create and sustain a competitive advantage in today's global marketplace.

Supplier Evaluation and Development

Because the emphasis regarding procurement has shifted to facilitating the company's manufacturing and marketing strategies with its supply base, it is important to develop the following perspectives when evaluating suppliers (Bowersox, Closs, & Cooper, 2007):

- **Continuous supply**—A core objective of procurement is to ensure that a continuous supply of materials, parts, and components is available to maintain operations. Stockouts of raw materials and component parts can cause production downtime and disrupt supply chain flows. This is often related to unexpected costs and inability to provide promised customer service levels.

- **Minimal inventory investment**—Another procurement goal is to achieve continuous supply with minimum inventory investment possible. The traditional approach to preventing downtime due to materials/parts shortage was to maintain a large amount of inventory. However, companies have realized that maintaining inventory is expensive and requires scarce resources. Thus, the ideal is to pursue a just-in-time strategy where needed materials/parts arrive just at the moment they are schedule to be used.

- **Quality improvement**—A commitment to continuous quality improvement in procurement is another critical perspective. Using poor-quality materials/parts, the quality of finished products or services will not meet customer requirements. Therefore, quality must remain as a focus of procurement.

The trend in supplier management is developing and maintaining good relationships with key suppliers. In contrast with traditional adversarial, low-price-centered buyer-supplier relationships, the cooperative, long-term partnerships with key suppliers facilitate information sharing and mutual commitment, thus achieving the optimal procurement performance.

Make: Producing Goods and Services

Operations focus on the production portion of the supply chain. This process refers to activities associated with the conversion of materials or creation of the content for services (Supply Chain Council, 2010). Manufacturing and service operations focus on the production of goods and services needed to fulfill customer requirements. Effective management of the production process and related resources is accomplished through planning, organizing, controlling, directing, motivating, and coordinating activities.

Role of Manufacturing and Service Operations in SCM

Operations functions closely interact with other supply chain functions. Both manufacturers and service providers need inventory support from their suppliers, and operations create the outputs that are distributed through supply chain networks. Production schedules must be coordinated with delivery schedules and transportation methods to ensure that inventory is received when promised. It is apparent that production operations are part of the supply chain and cannot be conducted in isolation. All activities in the procurement, production, and delivery processes need to be synchronized to ensure consistent, efficient product and service flows.

Key Elements of the Manufacturing Process

Production creates the outputs that are distributed through supply chain networks. Understanding the key elements of the production process helps supply chain managers improve product quality and customer responsiveness while achieving operational efficiency.

Capacity Planning and Scheduling

During the planning process, operations managers attempt to balance inputs, capacity (resources), and outputs in order to reduce waste. Both excess inputs and outputs create unnecessary inventory, and excess capacity will result in unnecessary production costs. However, shortages of inputs will deplete the production process and reduce output, and capacity shortages lead to overwork of machines and labor that may result in inferior quality of the finished product and service.

Two types of planning are particularly relevant for operations management: capacity planning and materials planning. Capacity planning is the process of determining the production capacity needed by an organization to meet changing demands for its products, whereas materials planning is the process of determining inputs that are needed for the production process. Operations planning occurs across three timeframes—long-range, intermediate-range, and short-range—and each has its own focuses. Long-range plans, covering a year or more, focus on major decisions regarding facility location, capacity determination, and aggregate production plans. Medium-range plans span 6 to 18 months and involve tactical decisions regarding identification of appropriate manpower and determination of inventory levels. Short-range plans, ranging from a few days to a few weeks, deal with specific issues and the details of production such as operational scheduling and job assignment.

Production Process Execution

Corporate strategies, available resources, and product characteristics all influence the execution of production operations. In terms of manufacturing strategies, several options can be considered. Selection is driven by the current state of the business environment, the need for the supply chain to manage demand variation, and the product's level of standardization and production complexity.

Make-to-stock (MTS) is the traditional production method where finished products are usually produced before the receipt of a customer order. With this approach, significant finished products inventory is typically manufactured in anticipation of future customer requirements. MTS makes production planning easier and more cost effective while enabling the manufacturer to offer short lead times. Accurate forecasting and inventory control are critical issues in MTS. This approach is most suitable for high-volume products with predictable demand or for seasonal products that require production in advance, such as certain stable consumer products, commodity-based end products (such as chemicals, pharmaceuticals, and energy products).

Make-to-order (MTO) is a production method that is triggered by an actual customer order or release, rather than a market forecast. For MTO products, more than 20 percent of the value-added takes place after the receipt of the order, and all necessary design and process documentation is available at the time of order receipt. The finished MTO product is generally a combination of standard components and options or accessories specified by the customer—often referred to as mass customization. Usually, the components are stocked in anticipation of customer orders, but the finished goods are not assembled until customer demand is known. MTO is useful in repetitive manufacturing situations where a large number of end products (based on the selection of options and accessories) can be assembled from common components. Automobiles and computers are good examples of MTO products that allow for limited consumer choice. Key benefits of this production method include better response time, reduced costs, lower inventory levels, reduced complexity from fewer items, and enhanced order entry speed and accuracy.

A build-to-order (BTO) production approach is similar to MTO in that it also delays assembly until a confirmed order is received for the product. However, it differs from MTO because of the higher level of customization and lower volume level of production involved. BTO is considered a good choice for uniquely configured products, such as airplanes. The main advantage of the BTO approach is its ability to handle variety and supply customers with the exact product specification required. Like ATO, BTO requires little or no finished goods inventory, which means carrying costs and product obsolescence rates are low. However, BTO faces some challenges, such as capacity utilization, which can be greatly affected by demand fluctuations, setup costs can be high, and lead times are relatively long.

Engineer-to-order (ETO), also referred as project manufacturing, is a method in which production focuses on the creation of highly tailored products for customers whose specifications require unique engineering design or significant customization. In this manufacturing environment, no two products are identical, and each order requires detailed cost estimates and tailored pricing. Each customer order results in a unique set of part numbers, bill of materials, and routings that tend to be complex with long lead times. Components and raw materials may be stocked but are not assembled into the finished good until a customer order is received and the product is designed. Successful ETO depends on close collaboration between all supply chain participants. Customers must be involved throughout the entire design and production process. Supplier engagement is also a critical aspect of ETO production. The materials required by the manufacturer can be very unique or ordered infrequently. ETO products include capital equipment, industrial machinery, and complex items in the aerospace and defense industries.

It must also be realized that these production strategies can often be used in combination. Delayed differentiation is a hybrid strategy that combines both MTS and MTO, which means a common product platform is built to stock and it is later differentiated by adding customer-specific features after the receipt of the customer order.

After making product strategy decisions, specific production methods should also be determined. Generally speaking, there are three main types of product methods: job production, batch production, and flow production.

Job production, also called jobbing or one-off production, involves producing custom work for a specific customer or a small batch of work. It is the oldest form of production. Individual products have few standardized parts. Job production is most often associated with classical craft production by small firms, but large firms also use job production (for example, building a new factory or installing machinery). Fabrication shops and machine shops whose work is primarily of the job production type are often called job shops.

Batch production is a manufacturing technique in which the product is produced stage by stage over a series of workstations. Batch processes resemble job shops, but with less variation in the routings among machines and larger batches. This process is useful to a firm that makes a number of different variations of similar products. Examples include bakeries or toothpaste manufacturers.

Flow production, also called continuous production, is a production method in which a product is continuously produced, flowing from one stage of production to the next. Workers and (increasingly) robots carry out individual repetitive tasks aiming to work as quickly as possible without loss of quality. This method is suitable for high-volume, undifferentiated products. Flow production systems are typically capital intensive, so it is important to keep them running smoothly with high levels of capacity.

Different production processes can be adopted to help production effectively meet the market demand. Depending on the processes chosen, the process layout is always an important decision to make, which involves the arrangement of machines, storage areas, and other resources within the four walls of a manufacturing or assembly facility. Common process layout concepts include these:

- A project layout is a fixed location layout where the product remains in place for the duration of production. Materials and labor are moved to this production site.

- A workcenter is a process-focused layout that groups together similar equipment or functions. The materials move from department to department for completion of similar activities and tasks.

- A manufacturing cell is another process-focused layout that dedicates production areas to a narrow range of products that are similar in processing requirements.

- An assembly line is a product-focused layout in which machines and workers are arranged according to the progressive sequence of operations need to make a product.

- Continuous process facilities are similar to assembly lines, but with its emphasis on the continuous rather than discrete nature of the product flow. This is an ideal method for producing highly standardized products such as gasoline, paper towels, and soft drink concentrate.

Effective selection of production processes can help an organization manage its variability of demand. Consistent demand patterns require far different manufacturing methods than do products whose demand is affected by the proliferation of competing products, short life cycles, and fluctuating demand. Companies must also establish facility layouts and production flows that are well matched to the demand volume and product variation.

Service Process Execution

Service companies differ from manufacturers in a number of ways, and some of the main differences between services and goods can be summarized as follows:

- Service cannot be inventoried. Its intangible nature determines that services are produced and consumed simultaneously.

- Services are often unique. Service customers often require high levels of customization.

- Services have high customer-provider interactions. In most cases, the service provision process involves the participation of the customer, whose inputs provide the necessary information for the server to provide needed services.

- Services are decentralized. Because services cannot be held in inventory and require high levels of customer interaction, the provider must be able to secure locations that are close to customers.

When purchasing services, customers actually buy a bundle of attributes, including the explicit service, the supporting facility where services are provided, and implicit services such the security provided or facility atmosphere. Successful service providers deliver this bundle of attributes in the most efficient way while still satisfying customer requirements.

Service delivery systems fall along a continuum with mass-produced, low-customer-contact systems on one end and highly customized, high-customer-contact on the other end. Many companies separate high-contact, front-of-the-house operations from low-contact, back-of-the-house operations to allow applications of different management techniques in order to maximize each area.

Service operations decision making involves several important factors. First, location is extremely important for most services because of its significant impact on customer visits, business volume, and profitability. Second, the service layout strategy should focus on reducing distance traveled (to reduce the travel time of customer or service provider's travel time between departments) and maximizing closeness desirability (for example, the cashier should be close to the entrance). Another key consideration is service capacity, which refers to the number of customers a company can service during a certain period. Most companies prefer less than maximum capacity to reduce the likelihood of generating long queues and waiting times. Often, queuing models will be used to maximize efficiency. Of course, service providers should also consider other relevant factors, such as employee training and technology application, because they all can directly affect the customer service level achieved.

Deliver: Fulfilling Customer Requirements

The deliver process focuses on the fulfillment of customer demand for the company's products and services (Gibson, 2005). Specifically, these are the activities associated with the creation, maintenance, and fulfillment of customer orders with the company's products and services (Supply Chain Council, 2010). Key logistics processes involving inventory management, distribution, and transportation are developed and executed to ensure the timely, low-cost flow of products to customers. These activities make for a very sizeable logistics industry. According to the U.S. Census Bureau (2012), the transportation and warehousing industry hired approximately 4.1 million employees in 2010.

Order fulfillment is a logistics element that is critical in SCM. No matter what forecasting accuracy level or product quality are achieved, unless the product can be properly delivered to the customer, the ultimate supply chain goal of serving customers cannot be accomplished. Furthermore, because the logistics and order fulfillment processes directly interacts with customers, it will greatly affect the customer's perception of supply chain performance. What is worth noting is that order fulfillment directly affects supply chain performance on the time dimension. Order fulfillment lead time is a widely accepted supply chain performance metric. CSCMP (2010) defines order fulfillment lead time as the average, consistently achieved lead-time from customer order origination to customer order receipt.

Effective management of the order fulfillment process also has important cost implications. For example, the more time required to process and prepare orders, the less time is available to deliver the order. This may force the company into the use of expensive expedited transportation. If the organization adopts technology to reduce the order transmittal to preparation time, then more time is available for transportation. This allows the company to use less expensive delivery methods and potentially consolidate deliveries for lower operational cost. The long-term cost-saving on transportation can also offset the investment on acquiring the new technology.

The logistics and order fulfillment processes involve three critical areas: warehousing, transportation, and order completion.

The Role of Warehousing in SCM

A warehouse has traditionally been viewed as a place to hold or store inventory. The focus of these warehouses has shifted from passive storage to strategic assortment and movement. Companies build strategically located warehouses to provide timely and economical inventory replenishment. Therefore, the term distribution center is used to capture this dynamic aspect of traditional warehousing.

In a perfect world, supply and demand would be balanced, with desired products being assembled on demand and delivered directly to the point of use. However, this goal is not feasible for most consumer products because production and consumption are not perfectly synchronized, transportation of individual units is too costly, and coordination of activity between such a large number of origin and destination points is very complex. To overcome such issues, distribution operations—distribution centers, warehouses, cross-docks, and retail stores—are established within the supply chain.

These inventory handling, storage, and processing facilities help supply chains create time and place utility. By positioning raw materials, components, and finished goods in production- and market-facing positions, goods are available when and where they are needed. Shorter lead times can be achieved, product availability increased, and delivery costs reduced, increasing both the effectiveness and efficiency of distribution operations. In highly contested markets, these responsive capabilities can help a supply chain enhance its competitive position. For example, Amazon.com spent almost $13.9 billion on fulfillment expenses—including opening 50 new warehouses during the period of 2010–2013. Such moves reflect Amazon.com's intention of competing with traditional retailers like Wal-Mart and Target in terms of customer order fulfillment (Kucera, 2013).

Enhanced customer service is not the sole rationale for inserting distribution operations into the supply chain. These facilities also help companies overcome challenges, support other processes, and take advantage of economies of scale. These roles include the following:

- **Balancing supply and demand**—Whether seasonal production must service year-round demand or year-round production is needed to meet seasonal demand, warehousing facilities can stockpile inventory to buffer supply and demand. Both situations require inventory storage to support marketing efforts. Storage provides an inventory buffer that allows production efficiencies within the constraints imposed by material sources and consumers.

- **Protecting against uncertainty**—Each supply chain faces various uncertainties, and warehouses can help companies better cope with these challenges. Warehousing facilities can hold inventory for protection against forecast errors, supply disruptions, and demand spikes. When forecasted demand is below actual demand or the market demand suddenly surges, the inventory held at the warehouses can meet these unexpected demand increases and maintain desired customer service levels. Also, warehouses' inventory can help preventing the possible disruptions caused by issues related to supplies, such as supplier shortage or poor quality supplies, and ensure continuous production flows. Achieving operations continuity is particularly important and challenging in today's time when many companies are relentlessly striving for efficiency.

- **Allowing quantity purchase discounts**—Suppliers often provide incentives to purchase product in larger quantities. To take advantage of such opportunities, companies often need to purchase more raw materials and parts than what they need immediately. In this situation, warehouses can handle the extra quantities, reducing the purchase cost per unit.

- **Supporting production requirements**—If a manufacturing operation can reduce costs via long production runs, or if outputs need to age or ripen (for example, wine, cheese, and fruit), the output can be warehoused prior to distribution. For manufacturers, strategic warehousing provides a way to reduce the holding or dwell time of materials and parts.

- **Just-In-Time (JIT) support**—Strategically located warehousing facilities support stockless production strategies. Centralizing a parts inventory at consolidated warehouses reduces the need for inventory at each production plant. Warehouses also become vital extensions of manufacturing through cross-docking, sorting, sequencing, and light assembly.

- **Promoting transportation economies**—Warehouses can also help companies achieve transportation-related economic benefits and efficiency. Warehouses have the ability to consolidate either or both inbound and outbound shipments. Larger-size shipments, often utilizing full-container capacity, generally result in lower transportation charges per unit than shipping partially loaded containers or moving small quantities at a time.

Warehousing facilities can provide numerous services, depending on the requirements of the supply chain. In general, four primary functions can be performed: accumulation, sortation, allocation, and assortment.

Accumulation involves receiving materials from a number of sources. The warehouse serves as a collection point for product coming from multiple locations and provides required transfer, storage, or processing services. The accumulation function allows organizations to consolidate orders and shipments for production and fulfillment processes. With accumulation there are fewer deliveries for customers to schedule and manage. Also, significant transportation cost-savings are achieved through larger, more cost-efficient deliveries.

Sortation involves bringing together similar products for storage in the warehouse or for transfer to customers. During the receiving process, goods are segmented according to their key characteristics—production lot number, stock-keeping unit (SKU) number, case pack size, and expiration date—and prepared for safe storage in the facility or immediate distribution. Proper sortation is essential for the effective management of inventory and fulfillment of customer orders. For example, mixing cases of fresh eggs with two different expiration dates on a single pallet can lead to improper inventory rotation and some product spoilage.

The allocation function focuses on matching available inventory to customer orders. The order is compared to inventory holdings, and available units are retrieved from storage according to the quantity ordered by the customer. This break-bulk capacity ensures product availability for multiple customers and allows them to purchase needed quantities rather than excess volume that is not desired.

Assortment involves the assembly of customer orders for multiple SKUs held in the warehouse. Such product mixing capability allows customers to order a variety of items from a single location in a single order. This avoids the expenses related to placing multiple orders and having them shipped from a number of locations.

In addition to these four key roles, warehousing facilities often perform other value-added functions and support evolving supply chain needs. Many warehouses are no longer simply viewed as places to store products, but as activity centers with flexible space and labor that can be utilized for a variety of customer needs ranging from product labeling to light manufacturing. These value-added activities help organizations handle special customer requirements, create supply chain efficiencies, and differentiate themselves from their competition.

Warehouse Execution and Control

Warehouses facilitate effective execution of product movement and storage, order fulfillment, and value-added services. The key warehouse processes fall into two categories: product-handling functions and support functions.

Product-Handling Functions

The primary warehousing activities focus on the movement and storage of product. Although storage is the more traditional and obvious function, maintaining proper product flows through efficient short-distance moves within the facility is a critical aspect of warehousing. Effective in-facility movement supports strong customer service and high inventory velocity, which reduces holding costs; lowers loss, damage, or obsolescence risks; and holds storage capacity requirements in check.

Product handling involves five primary processes: receiving and transferring goods into the facility from the transport network; putting away and moving goods into storage locations; order picking and selecting goods for customer orders; replenishing and moving product from storage locations to picking slots; and shipping and loading goods for delivery to the customer.

Support Functions

Although the product-handling functions account for the vast majority of the activity, labor, and cost in distribution facilities, a number of other administrative and management activities facilitate the successful execution of day-to-day operations. These support functions provide coordination between key processes and across the supply chain, protect the organization's inventory investment, and improve working conditions within the facility. Chief among these support functions are inventory control, safety, maintenance, sanitation, security, performance analysis, and information technology.

Together, these support functions facilitate product movement and storage within the warehousing operation and the fulfillment of orders. Without them, it would be difficult to protect workers and product from a host of challenges, maintain precise inventory records, or know how well the operation is performing. In short, the operation would quickly fall into disarray without the team of specialists behind the frontline managers and labor.

To achieve effective warehousing management, management must evaluate operational costs and productivity. Two primary categories of warehouse measurement are internal measures and customer-facing measures. Internal measures focus on the resources required to fulfill customer orders. These operational key performance indicators (KPIs) focus on asset utilization, labor productivity, and cost efficiency of the operation. Customer-facing measures focus on service quality issues and the ultimate goal of providing perfect order fulfillment. These KPIs also help to assess the reliability, responsiveness, and flexibility of the distribution operation. Both internally and externally focused KPIs are needed to evaluate the success and impact of a distribution operation.

Role of Transportation in SCM

A supply chain is a network of organizations that are separated by distance and time. The transportation is a key function that facilitates product flow between buyers and sellers, and allows organizations to extend the reach of their supply chains beyond local supplier capabilities and market demand. In essence, transportation involves the physical movement of goods between origin and destination points. The transportation system links geographically separated partners and facilities in a supply chain—customers, suppliers, channel members, plants, warehouses, and retail outlets—and it facilitates the creation of time and place utility in the supply chain.

The primary modes of transportation available to the logistics manager are truck, rail, air, water, and pipeline. In addition, intermodal transportation combines the use of two or more of the basic modes to move freight from its origin to the destination. Each mode has different economic and technical structures, and each can provide different qualities of link service.

Collectively, the U.S. transportation system moves approximately 12.5 billion tons of goods for businesses, valued at nearly $11.7 trillion (Bureau of Transportation Statistics, 2010). In terms of freight expense, organizations spent $760 billion for transportation services in 2010. Almost 78 percent of the total was spent on trucking services at $592 billion, which is followed by rail at 7.9 percent, air at 4.3 percent, water at 4.3 percent, forwarders at 4.2 percent, and pipeline at 1.3 percent (Coyle, Langley, Novack, & Gibson, 2013). The combined levels of freight value, volume, and spending suggest that truck, multimodal, and air transportation are premium-priced services for moving higher-value goods. In contrast, rail, water, and pipeline provide more economically priced services for lower-value commodities.

Companies have recognized the value of transportation is much greater than simply moving product from one location to another. Instead, transportation has been given strategic attention because of its role in SCM.

Transportation service availability is critical to demand fulfillment in the supply chain. No matter whether the transportation service is for moving goods between businesses or for delivering products to the end customer, reliable shipment is critical to ensure meeting customer demand. A transportation shortage would negate the organization's efforts to build and fulfill customer demand because the product inventory would not reach the stores in a timely manner. For example, Apple currently has suppliers worldwide and sells to customers all over the world from just a few plants. Transportation allows products to smoothly move across Apple's global network. Similarly, global transportation allows Wal-Mart to source products all over the world and to sell them to domestic consumers.

Transportation efficiency enhances the competitiveness of a supply chain. With efficient, effective transportation capabilities, organizations can build global supply chains that leverage low-cost sourcing opportunities and allow them to compete in new markets. In terms of supply management, cost-effective transportation enables companies to gain access to higher-quality, lower-priced materials and realize economies of scale in production. Likewise, low-cost transportation improves demand fulfillment opportunities. By keeping transportation expenses reasonable, the total landed cost of a product (its production costs plus transportation costs and related supply chain fulfillment costs) can be competitive in multiple markets. This is particularly true in today's environment in which many companies are sourcing from and marketing to overseas countries.

Transportation service capabilities are key to meeting customer requirements. High-quality, customer-focused transportation has a direct impact on a company's capability to provide the "Seven R's of Logistics"—getting the right product, to the right customer, in the right quantity, in the right condition, at the right place, at the right time, and at the right cost. In other words, transportation service capabilities must be aligned with customer requirements. In addition, transportation can create supply chain flexibility. By working with carriers that offer a range of transit times and service options, organizations can satisfy supply chain demands for expedited, next-day service as well as more economical, standard delivery requests. Many of today's companies are using existing transportation capabilities to provide innovative services to customers. For example, retailers such as Sears and Wal-Mart now offer free store pick-up service for online orders. Such services utilize existing store deliveries to help lower customers' purchasing cost.

Transportation cost is a critical driver of supply chain design and strategy. Accounting for nearly 63 percent of all expenditures for logistics activities, transportation spending far exceeds the amount of money spent on warehousing, inventory management, order processing, and other fulfillment system expenses combined (Wilson, 2013). Thus, transportation costs must be considered during the development of supply chain strategies and processes. In addition, transportation service availability, capacity, and costs influence decisions regarding the number and location of supply chain facilities. For example, many companies attempt to avoid locating distribution facilities in the state of Florida because little freight originates in the state. Carriers compensate for the empty outbound trips by charging higher rates to move freight into Florida. Also, transportation capabilities must align with the company's strategy. Amazon.com's decisions to provide same-day delivery and expansion into the fresh grocery business require low cost and timely transportation.

Given these critical roles, proactive management of transportation processes is fundamental to the efficient and economical operation of a company's supply chain. Company leadership must consider transportation issues when developing organizational plans, integrate transportation into supply chain processes, and optimize total supply chain cost rather than minimize transportation costs. Leading organizations like Best Buy, Amazon.com, and CVS have already moved in this direction. They recognize that supply chains can achieve time and place utility only through effective transportation processes that move goods to the place desired at the time required by their customers.

To better manage the transportation function and synchronize it with other supply chain activities, it is important to recognize and overcome potential obstacles, including supply chain complexity, competing goals among supply chain partners, changing customer requirements, and limited information availability. In addition, companies must cope with the dynamic business environment and trends discussed as follows.

Increased overseas operations pose significant transportation challenges. As companies expand their operations overseas to take advantage of lower costs or explore new markets, transportation processes must connect buyers and sellers that are thousands of miles apart. The vast distances associated with the extended supply chains produce not only higher transportation costs but also higher inventory levels caused by prolonged transit times and greater potential for supply chain disruptions. Therefore, companies must consider these higher delivery and inventory carrying costs when assessing the financial benefits of global sourcing and offshore manufacturing.

The demanding service requirements of customers affect transportation practices. Instead of shipping products in economic truckload or container load quantities, companies are forced to make smaller, more frequent deliveries to stay competitive. Shrinking order cycle requirements result in higher costs for faster delivery and longer fulfillment operation hours. In addition, customers' desire for real-time shipment visibility requires significant technological investments. Often, companies must align their operations with high-quality carriers that provide a balanced mix of capacity, speed, and consistency at a reasonable cost.

Transportation capacity constraints pose another significant challenge in moving freight across the supply chain. The current infrastructure has difficulty handling growing transportation volume. Congestion at ports and on highways has become the norm in many locations. A shortage of transportation labor makes it difficult for carriers to keep pace with freight growth. These capacity issues have caused higher freight rates, shipment delays, and limited negotiating capabilities. To overcome these issues, companies must take a more collaborative, flexible approach to working with carriers.

Rising freight rates are another major concern. After many years of low freight expenses fostered by excess carrier capacity, the situation has reversed. Capacity has shrunk due to greater demand for transportation services and industry consolidation from carrier mergers, acquisitions, and bankruptcies. The remaining carriers are now in a stronger position to increase rates to cover the rising costs of labor, insurance, and other expenses. They have also added fuel surcharges to their customers' freight bill. Although companies have limited control over rate increases and surcharges, they still need to proactively control transportation spending by using lower-cost modes of transportation when feasible, maximizing equipment utilization, and consolidating freight into larger shipments.

Transportation Execution and Control

When a shipment needs to be moved across the supply chain, transportation process must be carefully planned and executed. Decisions must be made regarding shipment size, route, and delivery method; freight documents must be prepared; in-transit problems must be resolved; and service quality must be monitored. The following are the common steps in transportation execution process.

1. **Shipment preparation**—Upon completion of the planning activities, it is time to prepare shipments for handoff to carriers for delivery. The transportation manager must choose the most appropriate carrier based on the size, service requirements, and destination of a particular shipment. The goal is to minimize transportation cost and protect the shipment while meeting service commitments. To ensure maximum effectiveness in the shipment-carrier matching process, many organizations use transportation routing guides. Because transportation provides final delivery to customers, an accurate freight count and a careful inspection should be performed prior to loading. During the loading process, the freight must be stacked properly and stabilized to minimize possible in-transit damages.

2. **Freight documentation**—Shipments are accompanied by related documents that facilitate the flow of goods. Specific documentation requirements depend on the origin and destination points, the mode being used, characteristics of the freight, and the carrier handling the freight. Common documents include the bill of lading that contains the terms and conditions of the transport contract, the freight bill that serves as the carrier's invoice for freight charges applicable to a shipment, freight claims form, commercial invoice, insurance certificate, and certificate of origin.

3. **Maintain in-transit visibility**—It is important to control the freight and manage key events as product moves across the supply chain. Visibility of in-transit freight is a key facilitator of this control because it provides the location and status of the shipments. Real-time shipment data makes it possible for companies to respond to problems in the transportation process. Various technologies such as satellite tracking capabilities have greatly enhanced the in-transit visibility.

4. **Monitor service quality**—Upon completion of freight delivery, transportation managers must analyze the outcome. Carrier performance should be carefully monitored with reference to the shipping contract. A popular approach for evaluating carrier service quality is to develop standardized scorecards or evaluation reports that prioritize the opportunities to improve supply chain performance.

Generally speaking, transportation metrics fall into two primary categories: service quality and efficiency. Service quality means doing things right the first time according to customer-defined requirements and expectations, such as consistent, on-time, and claim-free delivery. Efficiency focuses on transportation cost control. Key indicators of transportation efficiency include transportation cost per mile or per unit, asset utilization, and labor productivity.

Order Management

Effective order management is a key component of the logistics and order fulfillment process and is critical to operational efficiency and customer satisfaction. Conducting all activities relating to order management in a timely, accurate, and thorough manner can facilitate the effective coordination of company activities in other areas. Logistics decision making needs timely and accurate information related to individual customer orders. Thus, many companies place the order management function within the logistics area. In addition, the order management process significantly affects order cycle length, an important aspect of customer service. Hence, it is essential to have a better understanding of the customer order cycle process.

Key Elements of the Order Management Process

Traditionally, order management focused on the activities that occurred between order placement and receipt. However, many companies are adopting the order-to-cash (OTC) cycle concept, which includes all business activities in the order cycle plus the flow of funds back to the seller based on the invoice. The OTC concept has gained popularity because it accurately reflects the effectiveness of the order management process. During the OTC, order management activities coordinate all three major flows in a supply chain: products, information, and payment. The following sequential list provides an overview of the primary OCT major activities (Coyle et al., 2013):

1. **Order preparation**—In the initial step, the customer investigates product, pricing, and availability information to determine whether to place an order. The seller must have up-to-date information to respond quickly and accurately to the prospective buyer.

2. **Receive, enter, and validate order**—This step involves the placement and receipt of the order. This is typically accomplished through electronic data interchange (EDI), the Internet, or directly with a customer service representative who then enters the order into the seller's order management system.

3. **Reserve inventory and determine delivery date**—The seller's inventory levels are checked via the order management system to determine the best location for filling the order. After the inventory is reserved for the customer, the delivery date is set. The order is transmitted to the warehouse management system for fulfillment and to the financial system for invoice generation.

4. **Consolidate orders**—Orders are reviewed to determine opportunities for freight consolidation and batch warehouse picking. The goal is to identify order fulfillment cost efficiencies.

5. **Plan and build loads**—A delivery plan is developed around shipment size, delivery dates, and delivery method. An effort is made to optimize transportation cost within the constraints of customer delivery requirements.

6. **Route shipments**—The order is assigned to a specific route for delivery to the customer. Transportation management systems are often used to build efficient routes.

7. **Select carriers and rate shipments**—This step takes into consideration the size of shipment, destination, and delivery to determine the appropriate carrier and freight costs. Routing guides are established to facilitate consistent decision making.

8. **Receive product at warehouse**—The product is received at the distribution center. The order management system is checked to see whether the product is immediately needed to fill an open order. If so, the product is immediately combined with the on-hand inventory in preparation to be picked for the order. If not, it is stored in the warehouse for future use.

9. **Pick product**—Order fulfillment is scheduled and sequenced to optimize order picking efficiency while maintaining delivery schedules.

10. **Prepare and ship order**—The order is loaded, transportation paperwork is created, and the freight is released to the carrier for delivery. At this point, the seller invoices the customer.

11. **Customer delivery and receipt verification**—Once the shipment is delivered to the customer location, an inspection is completed to verify that the correct product and quantity was delivered. This step concludes the traditional order cycle. Successful completion of the order cycle facilitates faster payment from the customer.

12. **Install product**—If needed, installation takes place at the customer location. The success of the installation has an impact on customer satisfaction and the speed of cash flow back to the seller.

13. **Invoice**—Satisfied with product and order cycle performance, the buyer initiates payment to the seller. The final payment flow works its way back to the seller.

The length and reliability of the OTC cycle affect both the buyer and seller. When the OTC cycle time average is long or the variability is high, customers are forced to carry additional safety stock. This creates a cost burden for the customer and reduces satisfaction. In turn, the seller will have a difficult time maintaining a competitive advantage in the marketplace.

Return: Providing Post-Sales Support

The final supply chain process involves returns and other post-sale activities associated with the reverse flow of goods from customers to suppliers. The return process involves the customer request for returns, disposition decision making, scheduling of the return, and shipment and receipt of the returned goods. Post-sales support and customer relationship management (CRM) are also important supply chain activities. Organizations provide assistance after the sale through customer service, replacement parts, and repair services. CRM is the art and science of strategically positioning customers to enhance relationships and improve the profitability of the organization.

CRM and Customer Service Processes

Organizations have shifted their focus from reactively executing orders to proactively influencing how their customers place orders. The underlying rationale is that not all customers are equally profitable for the company. How customers order in terms of the products, quantity, and timing affect a company's cost of executing an order. The most profitable customers are those with ordering patterns that can maximize the efficiencies of the company. Using a CRM philosophy allows a company to identify and reward those customers.

There are four basic steps in the CRM process related to SCM:

1. **Segment customer base by profitability**—Companies use techniques such as activity-based costing to accurately allocate expenses to customers based on the specific costs of serving them. This involves analyzing how the orders are placed, the products involved, the quantities requested, and the timing. Normally, a cost-to-serve (CTS) model is used to evaluate each customer.

2. **Identify product/service packages for each customer segment**—Customer information and sales representatives' insights are used to determine what each customer segment values. Then, decisions are made about the value-adding products and services that will be offered to each customer segment. One solution is to offer the same product/service offering to each customer segment while varying the product quality or service levels. The other solution is to vary the product/service offerings for each customer segment.

3. **Develop and execute best processes**—This step seeks to deliver on the expectations identified. Companies often need to reengineer their processes to implement the differentiated product/service offerings.

4. **Measure performance and continuously improve**—Once the CRM program has been implemented, it must be evaluated to determine whether its goals are being achieved. Necessary adjustments should be made based on the outcome of the evaluation.

Although the concept behind CRM is simple—aligning the supplier's resources with its customers in a manner that increases customer satisfaction and supplier profits—the execution of a CRM program often presents significant challenges. CRM implementation usually requires changes in resource allocation, organizational structure, and market perception. For example, activity-based costing is a drastic deviation from traditional accounting approaches. It requires substantial resource investments, process redesign, and employee mindset changes.

Post-Sale Support

Although it is natural to focus on the goal of filling and delivering perfect orders, companies must realize that effective call center operations, service parts, and repair capabilities can also drive success. These post-sale service elements present a great opportunity for companies to distinguish themselves from the competition, especially when their products are undifferentiated. Superior warranty services, rapid fulfillment of spare parts orders, and quick repair capabilities have become core competencies of companies such as Caterpillar and Gulfstream.

The fast-growing online shopping and the shortening product life cycle have elevated the volume and importance of product returns. For example, Zappos.com has built easy and free return policies into its offerings as a service highlight. Also, companies such as GENCO have a core competency in managing electronics product returns and repairs for other companies.

Another important aspect of post-sale service is service recovery. No matter how well a company plans to provide excellent service, mistakes will occur. Contingency plans and procedures must be established for rapid service recovery to restore customer goodwill. Effective communication is an important component of the process to ensure that customers are quickly apprised of the situation and the solutions being implemented. Finally, employees are trained to diagnose service failures and empowered to quickly resolve them.

These post-sale services are necessary components of companies' service offerings. They help the organization build and maintain customer satisfaction, which evolves into loyalty and repeat business. Ignoring these post-sales activities and opportunities puts the organization at a competitive disadvantage and threatens its future growth prospects.

Trade-off Management

A challenging task for any supply chain professional is trade-off management. Every day, supply chain managers make many decisions to achieve optimized performance on different aspects of supply chain operations, such as speed, quality, cost, and others. However, these decisions rarely depend on a single factor or have a single outcome. Instead, they interact with each other and often create conflicting outcomes. For example, buying raw materials or parts in a large quantity to achieve purchase discounts will incur higher costs related to holding extra inventory. Also, using a slower transportation mode to reduce delivery cost will increase lead time and safety stock needs. Thus, compromises have to be made, and trade-off decisions occur daily across the supply chain.

Ideally, trade-offs are made with strategic forethought because they influence company performance and profits. The cross-functional and cross-organizational nature of SCM makes it very important to consider diverse stakeholder needs when making strategic trade-off decisions. Hence, they must replace rigid goals with responsive capabilities and cross-chain flexibility.

Trade-offs Between SCM Goals

Supply chain managers coordinate activities with internal departments and external partners, who often have incongruent requirements. Internally, production has quality goals, marketing wants effective customer service, and the finance department demands fiscal efficiency. Externally, product and service suppliers strive to maximize contract price while meeting performance standards. Customers want everything—quality products, perfect fulfillment, and rapid delivery—at a low cost. To balance these goals along with emerging sustainability requirements, supply chain managers must identify stakeholders' true needs versus their wants and then make decisions. This approach can help managers focus on appropriate process improvements, targeted collaboration, and sensible trade-offs.

Financial Trade-offs

Typically, supply chain managers follow an either-or perspective in which they pursue cost control with a limited focus on revenue growth or they might pursue revenue growth with rising costs. In this case, the strategic profit model can serve as a useful tool to evaluate the financial impacts of different options. Supply chain managers must think beyond the either-or trade-offs. Instead, they must simultaneously improve two or three financial goals through strategies such as process outsourcing. For a company without a core competency in SCM, using capable logistics services providers can reduce operating costs, increase revenues via improved service, and minimize capital investments.

Acquisition Trade-offs

Acquisition expenditures represent a major proportion of total supply chain costs. However, a narrow focus on unit purchase price with disregard to other priorities can lead to material shortages, expediting expenses, or product failures. Buyers must source goods at an optimal combination of cost, quality, and service after considering available purchase options and compromises:

- **Make versus buy**—Producing goods in-house generates greater control over production schedules and output quality, but increases capital costs. Outsourcing avoids capital costs and provides flexible capacity but sacrifices direct control and increases delivery cost and time.

- **Spot market versus contract purchasing**—Making spot market purchases creates sourcing flexibility but increases supply risks and price volatility. Contract purchasing generates capacity commitments from suppliers and stable prices but increases switching costs.

- **Strategic versus arms-length relationships**—Key supplier collaboration builds trust and leverages external expertise but adds overdependence risks. Using multiple suppliers as needed creates flexibility, provides multiple purchase options, and may reduce per unit costs; but these arms-length relationships may fail to create continuity or a continuous improvement focus.

Fulfillment Trade-offs

The fulfillment activities of inventory management, order processing, and transportation naturally create conflicting strategies. Narrowly focusing on specific efficiency or responsiveness goals in one area may lead to suboptimal overall performance. For example, relentless pursuit of lean inventories may create rush orders and expedited transportation will be needed to avoid stockouts. Providing perfect order fulfillment requires higher inventory levels and faster transportation than planned. To avoid these situations, supply chain managers must adopt a balanced perspective in which total fulfillment costs are controlled, subject to the organization's customer service commitments. This requires a cohesive fulfillment focus in which management focuses on customer satisfaction, total fulfillment cost, and thoughtful trade-offs.

Network Structure Trade-offs

Supply chain network design is a complex task that requires extensive knowledge of service standards, operating costs, and capacity requirements. It also often involves long-term capital investment and affects operating capabilities. Two important factors involved in network design include geographic scope and ownership.

From a geographic standpoint, decentralized networks can improve responsiveness and lower delivery costs through customer proximity but require greater facility investment and higher safety stock levels. In contrast, centralized networks expand the customer base and generate economies of scale via fewer facilities, but the fulfillment tends to be slower, costlier, and more complex.

Self-owned facilities provide control but require capital investment while leased facilities reduce the upfront investment and provide location flexibility but sacrifices control. Supply chain managers must understand this complex web of trade-offs to ensure that their network design decisions achieve desired outcomes.

How to Manage Supply Chain Trade-offs

Given the complexity in previously discussed trade-off decisions, it is important to keep in mind the following basic principles.

- Developing sophisticated trade-off tools that can rapidly estimate the benefits and drawbacks of strategic changes to the supply chain
- Creating robust cross-chain metrics that capture the full effect of supply chain trade-off decisions on multiple stakeholders
- Designing adaptable operating structures that support multiple solutions and a continuum of trade-off outcomes rather than rigid either/or trade-off goals

Effective trade-off decision making requires a reliable process to analyze complex interactions between processes and multiple organizations with conflicting objectives. A trade-off matrix supports analysis of situations with multiple solutions and objectives. The following process can be used to create the matrix:

1. List potential solutions.
2. List key objectives such as service, quality, and cost requirements.
3. Assign importance weights to each objective.
4. For each objective, rank the solutions from highest to lowest depending on their effectiveness.
5. Multiply each rank score by importance weight.
6. Add the scores across each row in a Total Benefit column.

This matrix process helps the organization analyze options and arrive at trade-off solutions with maximum value.

Technology plays an important role in supply chain trade-off analysis and decision making. Enterprise Resource Planning (ERP) systems, data capture technologies, and data warehouses help organizations rapidly amass accurate cross-chain data, which is vital for identifying improvement needs and trade-off options. Planning tools for network analysis, flow path optimization, and inventory right-sizing allow managers to test potential supply chain process changes long before implementation. Modeling, simulation, and scenario management packages support sensitivity analysis and risk assessment of trade-off alternatives. Execution systems for manufacturing, order processing, and transportation management help managers enact supply chain process changes and trade-off decisions. These tools help managers collect vital data, translate the data into information, use the information to drive trade-off decisions, and evaluate the decisions for intended outcomes. Proper integration of these tools promotes cross-chain optimization rather than functional suboptimization.

Another key consideration is the role of cross-chain metrics. People will act based on how they are measured. Supply chains are complex structures with multiple stakeholders who often have unique goals. Instead of using traditional KPIs that emphasize individual performances, it is important to develop cross-chain KPIs that expand functional boundaries. Such metrics are key to making sound supply chain trade-offs decisions and aligning different supply chain members' actions. Given enough time and executive support, these efforts will support collaborative supply chain trade-off initiatives and drive transformational supply chain improvement.

Successful implementation of supply chain trade-off decisions that avoid unintended consequences is not a simple task, but it is worthwhile. Organizations need the capability to articulate business level trade-offs that affect performance, analyze trade-off decisions with cross-chain metrics, and manage strategic trade-off decisions. By doing so, they will understand the potential positive and negative outcomes of trade-off decisions and be in a position to pursue the option with the strongest capabilities for creating cross-chain flexibility, responsiveness, and benefits.

Chapter Summary

Supply chains span multiple functions and organizations, making it necessary to adopt a business process perspective. This perspective and active alignment of key supply chain processes make up the starting point for accurate, timely, and cost-effective two-way flows of product, information, and payments between stakeholders. A strong and coordinated network structure provides the foundation for success.

As this chapter has highlighted, this is no easy task. Because supply chain control is spread among the organizations in the network and the goals may not always be consistent, it is imperative for the participants to reach a level of consensus regarding the five key supply chain processes: plan, source, make, move, and return. They must develop a clear and collective understanding of the role, elements, and execution of each process and clarify which organization will have primary responsibility for the process.

The chapter also indicates that these five processes are interdependent, connecting with and affecting each other on a daily basis. Supply chain managers must develop a holistic view of these processes and pursue strategies that optimize the whole chain versus the individual function or process. This requires effective trade-off management to ensure that managers within each process pursue overall supply chain success rather than independent priorities. A cross-chain perspective and global optimization goals underpin the principles and strategies in Chapter 3 that produce competitive advantage through supply chain excellence.

References

Andel, T. (1997) Information supply chain: Set and get your goals. *Transportation and Distribution*, 38(2), 33.

Bowersox, D. J., Closs, D. J., & Cooper, M. B. (2007) *Supply chain logistics management* (2nd ed). New York: McGraw-Hill.

Bureau of Transportation Statistics. *The Transportation Statistics Annual Report, 2010.* The U.S. Department of Transportation: Washington DC.

Cooper, M. C., Lambert D. M. and Pagh, J. D. (1997) Supply Chain Management: More Than a New Name for Logistics. *International Journal of Logistics Management*, 8(1), 1–14.

Council of Supply Chain Management Professionals. (2010) *Supply chain management terms and glossary*. Retrieved September 1, 2013, from http://cscmp.org/sites/default/files/user_uploads/resources/downloads/glossary.pdf.

Coyle, J. J., Langley, C. J., Novack, R. A., and Gibson, B. J. (2013) *Supply chain management: a logistics perspective,* Mason, OH: South-Western Cengage Learning.

Gibson, B. (2005, November/December) Piecing together the supply chain concept. *Blueprints: The Produce Professionals Journal*.

Hammer, M. (2001) The Super-efficient Company. *Harvard Business Review* 79(8), 82–91.

Harrison, A., & van Hoek, R. (2011) *Logistics management and strategy: Competing through the supply chain* (4th ed). Essex, UK: Pearson Education.

Kucera, D. (2013) Amazon ramps up $13.9 billion warehouse building spree. *Bloomburg.com*. Retrieved September 12, 2013, from http://www.bloomberg.com/news/2013-08-20/amazon-ramps-up-13-9-billion-warehouse-building-spree.html

Lambert, D.M. (2004) *Supply Chain Management: Processes, Partnerships, Performance.* The Supply Chain Management Institute: Ponte Vedra Beach, FL.

Mentzer, J. T. (2001) *Supply chain management.* Thousand Oaks, CA: Sage.

Mentzer, J. T. (2004) *Fundamentals of supply chain management: Twelve drivers of competitive advantages.* Thousand Oaks, CA: Sage.

Merriam-Webster Inc. *Merriam-Webster's Collegiate Dictionary*, Eleventh Edition. M-W Publishing: Springfield MA.

Supply Chain Council. (2010) *Supply chain operations reference model overview*: Version 10.0. Retrieved September 3, 2013, from http://supply-chain.org/f/SCOR-Overview-Web.pdf.

U.S. Census Bureau. (2012) *Statistical abstract of the United States: 2012.* Retrieved September 12, 2013, from http://www.census.gov/prod/2011pubs/12statab/trans.pdf.

Wilson, R. (2013) *24th state of logistics report.* Lombard, IL: Council of Supply Chain Management Professionals.

Wisner, J. D., Tan, K.-C., and Leong, G. K. (2011) *Principles of supply chain management: A balanced approach* (3rd ed). Independence, KY: Cengage Learning.

Zuckerman, A. (1998) The human side of information technology. *Supply Chain Management Review* 2(1), 80–86.

3

KEY STRATEGIC PRINCIPLES

Efficient and effective supply chains do not evolve naturally. A great deal of detailed planning, coordination, and negotiation is needed between stakeholders to create a supply chain with a logical structure, streamlined flows, and integrated processes. Reaching this level of supply chain coordination requires a deliberate and unified focus on leveraging beneficial supply chain principles. There must also be a conscientious effort to align supply chain strategies among the participating organizations. And awareness of potential supply chain barriers must be maintained. A failure to align participants along all three elements will lead to conflict and a fragmented supply chain that is incapable of reaching its potential.

Supply chain managers must also conscientiously prioritize principles and adopt strategies that align with the organization's goals. Alignment is necessary to ensure that the supply chain supports the financial success of the organization and maintains the patronage of C-level executives. Often, the organization's mission statement will provide strong direction for the development of supply chain strategic capabilities.

With a mission "to be Earth's most customer-centric company, where customers can find and discover anything they might want to buy online, and endeavors to offer its customers the lowest possible prices," Amazon.com (2013) must strive for a highly responsive yet cost-efficient supply chain. In response, the organization's supply chain leaders are rapidly building out worldwide supply chain fulfillment capabilities for its own customers and third-party sellers through a network of 94 distribution centers. This strategic spend of $13.9 billion will boost fulfillment facility automation and market proximity so that Amazon.com can serve its customers faster and more efficiently (Kucera, 2013). Ultimately, this supply chain strategy is well-aligned with the company's stated mission of being a customer-oriented company.

This chapter identifies the principles and strategies for establishing efficient, effective, and sustainable supply chains for organizations like Amazon.com that view supply chain management as a critical component of its success. The first section focuses on the key conceptual drivers and enablers of supply chain management. Next is a discussion of supply chain strategies that organizations employ to create value and sustainable supply chains. The chapter concludes with a review of the ever-present success barriers and challenges that can impede the effective execution of supply chain strategies.

By the end of this chapter, it will become evident that a supply chain without strategic underpinnings is not really a supply chain. Rather, it is a collection of disconnected activities that are highly susceptible to inefficiency, disruption, and failure. The only way to build a capable, competitive supply chain is to establish and execute strategies that are based upon the foundational principles of supply chain management.

Supply Chain Principles

A *principle* is a fundamental truth or proposition that serves as the foundation for a system (Oxford Dictionary, 2013a). Principles are important because they provide understanding of how things happen and why they happen the way they do. This knowledge is an essential tool like a compass and map that point decision makers in the right direction with their plans and actions. Though principles are general in nature, they drive the development of specific strategies, tactics, and execution processes toward logical decisions and successful results.

Guiding concepts, rules, and propositions have numerous applications across a field like supply chain management. They provide an important foundation for developing efficient and effective supply chain capabilities. For example, a principle such as the Pythagorean theorem or the distance formula provides the basis of the concept that the shortest distance between two points is a straight line. This principle underpins supply chain network planning and route design to determine optimal delivery paths in terms of cost, time, and constraints. Supply chain principles also help to create a consistent understanding of how supply chains work and contribute to organizational success. This drives stronger collaboration and coherent decision making between stakeholders.

In this section, 12 major principles of supply chain management are presented with specific attention devoted to their definition, importance, and use. They serve as important pillars of the supply chain concepts discussed in Chapter 1 and the processes explained in Chapter 2. These dozen principles are not the only ones followed by supply chain managers, but it would be difficult to achieve supply chain success without each one.

Demand Driven

Demand driven is the process of aligning the planning, procurement, and replenishment processes to actual consumption and consumer demand. Typically, the retailer possesses good demand information because they sell directly to the consumer. This current consumer sales data, also known as point-of-sale (POS) data, helps retail organizations to better understand consumer demand trends. The insight gained by POS data allows retailers to perform relatively accurate demand forecasts. However, the further removed that each supply chain partner is from the consumer, typically the worse the forecasting accuracy.

Organizations further removed from the consumer may not have access to POS data. Although each supply chain network entity may consult their business partner's demand forecasts, often each firm performs its own independent forecast. The independent forecasting process can introduce additional uncertainty and variation into the results. Therefore, the error rate of the forecast tends to increase with each echelon of the supply chain that is further removed from the consumer. As demand forecast accuracy diminishes, supply chain performance and customer service levels decline.

Supply chain network partners have responded by sharing POS data and demand forecasts. Often this process starts with the retailer sharing valuable POS data with supply chain partners. This allows supply chain network participants to benefit because everyone has access to the same POS data.

Sharing POS data has resulted in several beneficial supply chain management practices. For example, many organizations now partner with each other to use POS data to inform the Collaborative Planning, Forecasting, and Replenishment (CPFR) process. This process consists of supply chain network partners who collaborate to establish a joint forecast and then perform collaborative planning and replenishment activities.

Vendor Managed Inventory (VMI) is another common strategy enhanced by sharing POS data. Many retailers, such as Wal-Mart, Target, and Lowe's, share POS data with suppliers. The supplier becomes responsible for monitoring store inventory levels and managing the fulfillment process for every product they supply to the retailer.

VMI is successful in part because vendors benefit when they prevent retailer stockouts. Stockout prevention minimizes lost sales of the supplier's products, enhancing both supplier and retailer revenues. Another advantage of VMI is efficiency. The retailer has thousands of store items to manage, enhancing restocking error potential. The vendor has comparatively fewer items to monitor and replenish, reducing error potential.

Total Cost Focus

A *total cost focus* refers to achieving an understanding of total landed costs and the total cost of ownership. Having a total cost focus is important for many functional areas of the supply chain. However, perhaps nowhere is this focus more important than in the acquisition process.

Visualize a supply chain with the consumer at one end. It is likely that located at the opposite end is some form of raw materials supplier. Raw materials and component parts suppliers are usually several echelons from the consumer. Nevertheless, suppliers play a critical role in the supply chain because they provide key inputs that set the transformation process from a raw material to a finished product in motion.

When acquiring items, there are many issues to consider beyond the cost per unit. A potential supplier may offer the lowest cost per unit and claim to provide overall cost savings. However, the anticipated cost-savings may not translate into the lowest total cost when all factors are considered.

A buyer must evaluate the total cost of item acquisition. This goes beyond merely the cost per unit to include additional variables. What is the total landed cost of the item to your facility? Perhaps there is a significant increase in the logistics costs to deliver the item from the supplier to your manufacturing facility. In this case, although the cost per unit may be lower, the overall "landed" cost of the item may actually increase.

Total cost analysis should encompass the total cost of ownership of an item over its entire lifespan. This type of analysis evaluates the total cost from the cradle to the grave. Often these types of analyses even include future disposal costs.

Beyond the direct item costs, the organization's operational effectiveness may also be affected (Chopra & Meindl, 2013). For example, consider a supplier located a significant distance from a buyer's manufacturing facility. The risk of a supply disruption tends to increase with distance. The disruption may result a failure to meet a predetermined delivery schedule, creating a costly disruption in the manufacturing process. Although not a direct acquisition cost, these variables should be evaluated as part of the overall supply management decision-making process.

Consider the impact of an item failing to meet predetermined quality standards. The total cost of item ownership can increase rapidly. If the quality issue is detected, it must be resolved. The resolution takes time and resources, reducing operational efficiency. This may delay manufacturing and alter the production schedule.

If the quality issue is not detected, the item is incorporated into the finished good, possibly negatively affecting functionality and quality. This can negatively affect sales revenues, warranty costs, and customer satisfaction. In this case, although the cost per unit was low, the total cost of ownership was high. Many trade-offs must be evaluated if the organization is to achieve the lowest total cost of ownership.

Segmentation

Segmentation is the ability to effectively identify critical requirements of key groups of customers and develop service packages designed to provide significant value to each group. Although no two consumers are identical, many have common wants, needs, and desires. As a result, you can segment consumers into a few key segments of consumers who have similar traits, buying habits, or service requirements.

Why segment consumers? Simple; from the view of the firms servicing consumers, not all are equal. Some consumers are high-volume, high-profit customers; others infrequently purchase low-profit items. It would be unwise for an organization to treat all customers as if they contributed equally to the company.

Supply chain planning focuses on how to best serve the key customer segments of an organization. To properly serve a diverse set of consumers, network planners must develop a thorough understanding of their customers. This includes understanding the common traits of consumers so they can segment them into a few key groups with similar supply chain service requirements.

One popular segmentation strategy is to establish similar groups of customers and then differentiate service levels by customer group. One way to accomplish this is through ABC analysis. ABC analysis categorizes customers by key variables of importance to the organization. Customers can be segmented by many attributes. Some of the more commonly used measures include revenue, profit, volume, and purchase frequency.

The tiered frequent flyer program of your favorite airline is an example of customer segmentation. Typically, airlines segment customers based on usage frequency or distance traveled. Customers who achieve the highest tier of the frequent flyer program are labeled A-level customers and are rewarded with certain benefits such as free upgrades and preferred boarding.

Many companies use segmentation for many different purposes. Restaurants have frequent visitor programs and grocery stores reward high-volume consumers with coupons. Although these types of programs are popular, there are many other uses of consumer segmentation information.

An electronics company may offer A-level customers unlimited free technology support while C-level customers receive 3 hours of free technology support before they are charged for the service. A mail order organization may offer free overnight shipping to their most critical, A-level customers but limit free shipping to the 3-day ground transportation option for B-level customers.

Armed with this type of segmentation information, planners can design supply chain networks capable of servicing the needs of each primary consumer segment. This allows network designers to establish tailored service capabilities specifically designed to meet the needs of a specific targeted group of consumers. In extreme cases, customized supply chain services may even be offered to those few critical customers who are vital to the company.

Customization

Customization is the ability to adapt a supply chain network to meet the critical service needs of key customer segments. Customization builds upon the previous discussion about segmentation. Different customer segments have different service requirements. However, many organizations have traditionally designed one supply chain network to service all customers. In the best case, this type of network may adequately fill the service requirements of some customers or segments. Worst case, this approach may result in a network that doesn't adequately fill the service requirements of any customers or segments.

Consider the automotive tire manufacturer who must simultaneously service large automotive manufacturers such as Ford and General Motors, the large tire retailers, and the smaller local tire retailers. The demand pattern for each customer segment is likely to be quite different. As a result, each customer segment's service requirements are likely to be different. This could necessitate three different distribution network designs.

Perhaps the large automotive manufacturers require frequent large shipments of a single brand of tire. These shipments may be characterized by limited product variety and all delivered to one manufacturing location. Conversely, a large retail operation may require less frequent, large shipments of multiple brands of tires. These shipments are characterized by significant product and size variety. Perhaps these shipments need to arrive at eight different regional distribution centers at variable time intervals. Contrast these delivery scenarios with the small local garage selling tires to local customers. Perhaps the local garage requires frequent shipments of a limited number of tires with a specific brand and defined specifications.

These scenarios create challenges for any supply chain planner. In addition to limited resources and the desire to adequately service all customers, network efficiency remains a key consideration. Network efficiency is usually aided by the development of a single standardized supply chain network. However, network standardization may not coincide with adequately servicing the needs of each customer segment. Faced with this conundrum, many companies now turn to a third-party logistics (3PL) provider to help improve network flexibility and capacity.

External expertise often creates opportunities not available to the entity faced with meeting the needs of their customer segments. Not only can the 3PL bring expertise and a fresh perspective but it also have other customers. Often this can create opportunities for efficiency through consolidation while simultaneously enhancing supply chain flexibility. This may assist a company in meeting the diverse needs of its customer segments in a cost-effective and efficient manner.

Postponement

Postponement is a concept whereby activities in the supply chain are delayed until actual demand is known or realized (Van Hoek, 2001). Postponement can appear in many forms and can be applied to upstream, downstream, and distribution operations (Waller, Dobholkar, & Gentry, 2000). If designed and implemented properly, postponement has the potential to positively affect many areas of operations throughout the supply chain (Boone, Craighead, & Hanna, 2007).

Consider a customer entering a paint store to select a color of paint for their kitchen. The customer is greeted with thousands of color choices. Have you ever thought of the difficulties of stocking thousands of colors at each paint store? What happens to the non-purchased paint that must be disposed of due to spoilage or age?

Retail paint stores avoid these complications by utilizing postponement. Store personnel wait until the customer enters the store and selects a specific type and color of paint. Store personnel respond to the customer by retrieving a can of white paint from inventory and adding the appropriate amount of dye to create the color the customer desires. The store uses postponement until it can configure the product to the order.

The postponement strategy eliminates the need for the paint store to forecast demand for specific colors of paint. Instead, it forecasts demand for paint and then uses a configure-to-order strategy once demand is known. This is a much simpler forecasting proposition. Furthermore, postponement enhances customer responsiveness through product customization while also reducing unnecessary waste associated with spoiled inventory.

Postponement is also implemented early in the supply chain process. Postponement can provide the opportunity to delay raw material orders until customer orders are received. This drastically reduces the acquisition of unnecessary inventory. Furthermore, the efficiency of the supplier manufacturing process can be enhanced because of an improved understanding of and response to demand. Postponement's potential to create such significant supply chain improvement has not gone unnoticed by practitioners.

Common industry examples of postponement include Benetton and Hewlett Packard. Benetton used postponement to improve its responsiveness to customer demands. By postponing the dyeing of its garments, Benetton is better positioned to respond to demands for popular colored clothing and reduce inventory of less-popular colors (Dapiran, 1992). Similarly, Hewlett Packard has postponed final assembly of its products until the late stages of the supply chain. The postponement of final assembly coupled with the shift of assembly locations closer to customers has resulted in a more cost-efficient production process while reducing transportation and logistics costs (Feitzinger & Lee, 1997).

Visibility

Visibility is the use of technology to create real-time knowledge of cross-chain activities. There is a popular saying that knowledge is power. In today's supply chain environment, information is necessary for knowledge. Given today's technology, significant amounts of information are generated and collected. However, the information is useless until it is turned into knowledge and used for informed decision making.

In today's global supply chain environment, products often move thousands of miles through a number of countries and are handled by numerous people representing different entities of a supply chain network (Coyle, Langley, Gibson, & Novack, 2013). A byproduct of this type of system is that, in spite of the best efforts to maintain product velocity throughout the supply chain, network disruptions periodically occur. Ideally, the supply chain network is designed to anticipate and avoid many potential disruptive events.

However, in the event a disruption occurs, having network-wide supply chain visibility becomes critical. Visibility informs the decision-making process and assists with rapid recovery to help to minimize the impact of the disruption (Simchi-Levi, Peruvankal, Mulani, Read, & Ferreira, 2007). This early warning system is significantly enhanced by the creation of visibility throughout the supply chain.

Today's marketplace is characterized by significant technological advances. These scientific advances have led to a plethora of information technology-based solutions. Nevertheless, supply chain wide visibility remains a significant challenge for most organizations.

Supply chain networks are typically made up of at least several proprietary companies, each operating its own systems and tracking its own information. Although not impossible, attempts to bridge the communications and information gaps between organizations of a supply chain network have often limited the ability to achieve supply chain-wide visibility. Different technologies, software systems, and computer capabilities all spanning numerous proprietary organizations across the globe have limited the effectiveness of this initiative in many situations.

When supply chain-wide visibility is achieved, the benefits can be enormous. The day-to-day operational aspects of the supply chain network are enhanced. Network-wide planning and decision making are aided by information sharing and the creation of supply chain decision support systems. System-wide strategic analysis is enhanced and collaborative initiatives become possible.

The key to supply chain-wide visibility begins with technology and information system integration. Once integration is achieved, network-wide connectivity creates transparency, enhances information sharing, and facilitates events management and control throughout the entire supply chain network. Once achieved, visibility allows organizations operating within the network to make strategic adjustments to enhance supply chain performance and create customer value.

Evaluation

Evaluation is the use of a metrics system to provide valuable information that is useful for managerial decision making. It is difficult to manage what you can't measure. Therefore, measurement is a key aspect of an effective evaluation system.

Supply chain planners focus a significant amount of attention on meeting customer service requirements. But how does an organization really know if it is meeting customer service expectations? How can an organization make adjustments to improve supply chain network performance if it is not aware of any shortcomings of the current network?

Most organizations have evaluation mechanisms in place to measure many different types of organizational performance, including personnel, logistics, finance, marketing, operations, and others. Although there are numerous supply chain-oriented metrics, some of the more common may include on-time performance, perfect order percent, inventory turns, and product defect rates. However, although these types of metrics are common to many organization's evaluation processes, often supply chain-wide performance is not well measured.

Organizations of a supply chain network that embark on the development of a framework for system-wide evaluation must identify, agree upon, and adopt channel-spanning metrics or key performance indicators (KPIs). The channel-spanning metrics must have common definitions and be measured in a common manner if the evaluation process is to be effective and produce information useful for decision making. Although the adoption of a system-wide evaluation process does not prevent each organization from measuring other metrics or performance indicators, certain core metrics must be common across the entire network.

An evaluation system based on common metrics or KPIs is useful in part because of its ability to help monitor network performance. A KPI with an unacceptable reading may be an indicator of a needed network adjustment. Therefore, many organizations display KPIs in the form of a scorecard or dashboard, making it easy to monitor network performance. Similar to gauges of your automobile, a metric displayed on a dashboard provides valuable information. If the metric is out of tolerance, it may be an indication of a problem.

Consider the temperature gauge on your automobile dashboard. If engine temperature becomes excessive, the gauge will produce an out of tolerance reading, warning the driver that the engine is not functioning properly. The gauge does not specifically identify the cause of the excessive temperature; it merely alerts the driver of the malfunction. It is necessary to explore the situation further to determine the cause of the elevated temperature.

Network-wide metrics operate in much the same manner. An organization operating within the supply chain network may discover a KPI that is out of tolerance. The KPI may show that velocity of product through the distribution center is unacceptable. Once discovered, the situation must be investigated to identify and correct the cause of the problem.

Agility

Agility refers to improving the responsiveness of a supply chain to market, demand, and supply volatility while delivering the same or comparable cost, quality, and service. In today's environment, it is necessary to have an agile supply chain. Customer demand is affected by numerous variables and therefore tends to fluctuate and be difficult to accurately predict. Nevertheless, customer demand ultimately drives the entire supply chain process.

The operating environment of today's supply chain networks is one where customers expect almost instantaneous service. However, often firms require significant lead time to properly prepare the product for consumers. These companies must rely on their ability to quickly respond to changes in demand.

Responding to a marketplace characterized by constant demand fluctuations has led many organizations to reduce product variety by implementing mass customization principles. For example, a company manufacturing several varieties of durable household appliances may delay final subassembly until the product travels through the majority of the supply chain. Once a specific customer order is received, final subassembly is performed.

For example, Whirlpool achieved significant cost-savings by shipping standardized appliances to retail stores. Once a customer orders a particular model, the correct control panel is installed. This allows Whirlpool to quickly respond to specific customer demand. Remaining agile was critical to achieving a significant reduction in inventory and transportation costs (Waller et al., 2000).

At any one time, billions of dollars of product is moving around the globe through numerous, intricate supply chain networks. Today's complex supply chain networks are designed to facilitate the movement of inventory. Inventory is a costly investment, leading companies to develop a systems approach that maximizes the velocity and availability of inventory throughout the supply chain (Boone, Craighead, Hanna, & Nair, 2013).

Supply chain disruptions do occur and can be costly. A disruption can increase stockout rates, delay customer deliveries, alter production schedules, and create a ripple effect that can cause costly plant shutdowns. The quicker a disruptive event is discovered, the more time is available to proactively manage the situation to minimize or eliminate the negative effects of the disruption.

Having an agile supply chain network capable of quickly adjusting to the disrupted environment and minimizing its impact is extremely important in today's globally competitive environment. Firms are finding that agility tends to be aided by several factors, including simplifying the network when possible. Common practices include removing unnecessary network echelons, automating product flows, and replacing inventory with information used for decision making.

Integration

Integration is the effective alignment of network partners through incentives designed to enhance supply chain performance. As customer service expectations and competition levels rise, individual firms and entire supply chain networks search for ways to differentiate themselves from competitors. However, if network partners are not working together in an integrated manner, differentiation in an intensely competitive marketplace is nearly impossible.

One potential positive differentiator is to offer customers unique supply chain services. Each customer segment is different and has unique service requirements. These differences may lead to multiple variations of a supply chain network. Occasionally, network adjustments must be made to serve the specific needs of each customer segment.

Offering unique supply chain services can differentiate participants of one supply chain network from those of another network. However, this may require some customization of the distribution network. Successful customization of the distribution process is accomplished when each specific customer segment's service requirements are met. Typically, this requires supply chain-wide participation, which is nearly impossible without complete integration of the partners of the supply chain network.

Integration of all participant firms requires an alignment of common goals, objectives, and incentives. Different organizations of a supply chain network fill different roles. Therefore, different organizations may focus on achieving different goals and objectives. Although not unexpected, all supply chain partners must align themselves sufficiently to pursue the same general goals and objectives.

What is critical to one organization may be relatively less important to a network partner. Therefore, incentive structures of firms tend to vary, creating an environment of inconsistency. One firm's personnel may be rewarded for pursuing a particular objective, whereas personnel of a network partner may be encouraged to pursue a different objective. This is a common reason why supply chain-wide integration is difficult to achieve.

Integration helps to align the interests of all the firms participating in the supply chain network (Lee, 2004). This is critical if firms are going to team up to differentiate their supply chain network service from those of competitors. Effective system-wide integration aligns the goals of each firm within the network to the goals of the overall network. The result is a situation in which the optimal benefit to the firm matches the optimal benefit for the entire supply chain network.

Efficiency

Efficiency refers to reducing the costs of supply chain processes through a concerted effort to reduce errors, improve productivity, and enhance asset utilization. Maximizing supply chain process efficiency is critical to success in today's highly competitive marketplace.

In fact, many companies find themselves in a position where their supply chain network must compete against other networks for customers. This complicates the operating environment when compared to the more traditional competitive structure where a company focuses on direct competition with a few firm rivals.

Efficiency tends to be aided by lean practices throughout the supply chain network. Lean initiatives can be applied to many product and service-based functional areas of the supply chain. Lean initiatives can be responsible for reduced inventory levels or improved carrier service levels. However, for lean initiatives to be fully realized, access to accurate and timely information is necessary.

Consider a retail customer expecting a critical Wednesday shipment from a supplier. When the shipment fails to arrive, the retailer informs the supplier. Due to supply chain network limitations, the supplier is incapable of accessing information, allowing them to monitor inventory positions throughout the supply chain. Without knowledge of the original shipment status, the supplier incurs considerable expense to prepare a second shipment for next day delivery to the retailer.

If the supplier had visibility into the transportation provider's information system, they would have discovered that the original shipment was delivered as promised. The shipment was not properly entered into the retailer's inventory system, so the retailer was unaware the shipment had arrived on time. If the supplier had access to the carrier's information system, it would have known the exact time and location of delivery and could have informed the retailer of who signed for the shipment and when it was received.

The previous example is one of many ways supply chain efficiency can be enhanced. Another frequently used tool is optimization. Carriers can use optimization to enhance transportation routing and scheduling. This improves network efficiency by enhancing asset utilization. Inventory managers use optimization to enhance asset utilization and improve network efficiency, which is accomplished by evaluating and repositioning valuable inventories. Similarly, supply chain designers can use simulation methods to analyze different network designs and develop optimal networks.

Efficiency is a key attribute to a successful supply chain network that should be rigorously pursued. In fact, many supply chain managers have implemented a continuous improvement process in pursuit of efficiency. Continuous dedication to the pursuit of efficiency is important because supply chain strategies, market demands, competitors, technologies, and customer expectations are constantly changing.

Resiliency

Resiliency is the development of an ability to sense and respond effectively to a variety of incidents and crises. Customers anticipate perfect deliveries that are void of any interruptions. Unfortunately, supply chain disruptions do occur. In these instances, supply chain managers must have a contingency plan in place to minimize the impact of the unexpected event on network performance.

In the event of a disruption, customers still expect end-to-end supply chain service complete with a resilient, visible network that responds rapidly to unexpected issues. Although customer expectations remain high when a disruption occurs, given today's highly competitive marketplace, supply chain participants also pursue network efficiency to remain competitive. One component of achieving network efficiency involves a lean network. However, pursuing a lean, efficient supply chain can further compound the challenges of designing and implementing a resilient network.

Consider a lean manufacturing operation utilizing Just-In-Time (JIT) principles to manage their component parts inventory. To pursue efficiency, the manufacturing operation holds 2 hours of inventory. Under normal operating conditions, this type of JIT inventory practice positively affects supply chain efficiency. However, these types of lean practices can further elevate the risk caused by a disruption and increase the recovery costs associated with an unexpected event.

In the event of a disruption, the lack of safety stock adds to the importance of having a resilient supply chain network. When an event negatively affects the supply chain by interrupting normal product flow, one common consequence is the occurrence of delivery delays. Under these circumstances, the 2 hours of safety stock previously contributing to supply chain efficiency now threatens to contribute to a plant shutdown. However, a resilient supply chain with built-in safeguards to help minimize or eliminate the negative effects of a disruption may be able to help prevent a costly plant shutdown.

The design of a resilient supply chain network starts when all supply chain participants continually identify, measure, and evaluate their operating environment. Conducting a continual risk assessment process is one of the necessary components of designing a resilient supply chain. Another precursor to a resilient supply chain is the use of risk assessment results to develop the ability to sense and respond to incidents and crises in an effective manner.

A popular preemptive solution to a potential disruption is the establishment of a contingency planning process. Performed properly, contingency planning enables an organization or supply chain network to be more effective in their prevention of, and response to, a disruption. Contingency planning is actually a risk management technique that contains many of the attributes that are widely used to promote organizational or network resiliency.

Sustainability

Sustainability is the design and improvement of supply chain networks by embedding strategies and operational practices that reduce the environmental impact, resource waste, and cost of operations. The importance of sustainability initiatives has grown considerably over the last decade. More than ever before, customers are holding companies and their supply chain partners responsible for a network that is both effective and environmentally friendly.

All of the functional areas of a supply chain can implement initiatives to enhance network wide sustainability. Packaging is one popular area many suppliers and finished goods manufacturers are evaluating to improve sustainability. Although always important, packaging is critical when products must navigate a complex network composed of several supply chain partners.

Packaging provides product protection and serves to promote goods and convey valuable information to consumers. Packaging is necessary to maintain product integrity. However, once the packaged goods are consumed, the packaging typically becomes useless and must be discarded. In response, companies now review and update packaging procedures to reduce waste and minimize any negative environmental impact from packaging materials disposal.

As products navigate today's complex supply chain networks, transportation becomes a critical functional area of the system. Over the last decade, many transportation companies have focused on reducing their carbon footprint. Carriers utilize cross-docking facilities, consolidate shipments, improve routing and scheduling practices, and implement fuel-saving strategies to simultaneously enhance sustainability and reduce costs.

Given the heightened environmental awareness of customers, achieving a competitive advantage through sustainability is beneficial to establishing a positive company or network perception among customers. In fact, consumer awareness of environmental issues is so prominent that many supply chain network participants have implemented sustainability requirements for their business partners. In many cases, these companies enforce supply chain partner compliance through regular company audits. Significant scrutiny from regulators and cost reduction are also powerful reasons for transportation providers to pursue sustainability initiatives.

Although packaging and transportation are two common examples of sustainability initiatives in a supply chain, there are tremendous opportunities for all functional areas of the network to contribute to environmentally friendly initiatives. Office functions can reduce paper usage, save on copy toner, and reduce energy consumption. Warehouses can implement warehouse management systems designed to improve operational efficiency, saving fuel for equipment and improving product picking productivity. Suppliers and manufacturers can implement reusable container initiatives and recycle pallets and packaging materials. Overall, there are tremendous benefits to be gained by a well-designed and implemented supply chain network-wide sustainability program.

Supply Chain Strategy

A *strategy* is a plan of action or policy designed to achieve a major or overall aim (Oxford Dictionary, 2013). The military perspective of strategy focuses on the planning and directing of overall military operations and movements in a war or battle. In business and supply chains, the primary aim is to create sustainable competitive advantage. As such, strategy drives managers to define the organization's position, make trade-offs, and forge fit among activities (Porter, 1996).

Strategy is critical to supply chain management because it defines and represents the organization's commitment to pursue a particular set of actions. A supply chain strategy determines the broad structure of the supply chain and the nature of materials procurement, inbound and outbound transportation, production, distribution of products to customers, and follow-up services. Supply chain strategy also specifies design decisions for these processes, whether these processes will be performed in-house or outsourced, and what the processes must do particularly well (Chopra & Meindl, 2013).

Ultimately, supply chain strategy focuses on the *how* of a supply chain: how it will operate, how it will be managed, and how performance will be improved. These *how* considerations generate a set of actions that create the capabilities the organization needs for the future. That is, the strategy must analyze all the critical inputs listed in Figure 3-1, determine the critical issues to address, define a strategy to deal with these critical issues, and then identify a set of action plans to develop new and enhanced capabilities (Dittmann, 2012).

As few as 15 percent of companies may have a documented supply chain strategy in place (Slone, Dittman, & Mentzer, 2010). Given this, the creation of a supply chain strategy is likely to provide a competitive benefit to the creating firm. In this section, 12 elements of supply chain strategy are discussed with specific attention devoted to their meaning, importance, and application. Each can support operational effectiveness and provide a strong ability to perform the supply chain processes discussed in Chapter 2 better than the competition. More important, they can enhance the organization's ability to perform similar activities in different ways from the competition or different activities from the competition. This is the basis of strong strategic positioning and sustainable advantage (Porter, 1996).

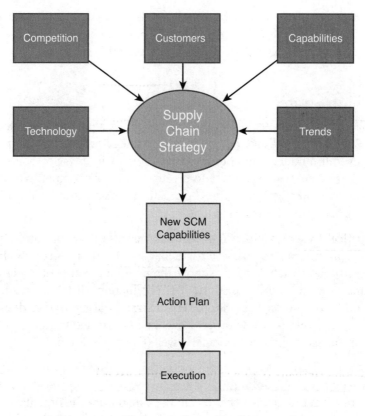

Figure 3-1   Elements of a supply chain strategy

Demand Sensing and Responsive Strategy

The old adage states, "One constant in business is change." Many of the best-performing supply chains have taken heed of this adage and developed the ability to rapidly adjust their supply chain processes, allowing them to remain in a constant state of preparedness. *Demand sensing* is defined as shortening the time to sense true market data to understand true market shifts in demand response (Cecere, 2013). A demand sensing and responsive supply chain strategy actually consists of two distinct but complementary components. Demand sensing provides the capability to detect changes in the customer, competitive, and technology environments that may eventually lead to changes in demand. The complementary responsiveness capability allows the organization to quickly adapt to the changes as they occur. A demand sensing and responsive strategy has also been referred to as *adaptability* (Lee, 2004).

Demand sensing capabilities should be forward-looking, focused on just forming market trends and broad economic shifts. Clearly this requires more than simple data collection and relies heavily on interpretation of subtle trends and incomplete information. Done correctly, however, the company benefits from identifying early signals of potential major shifts in demand, and potential new regional opportunities where low-cost operations may become viable. The most responsive companies are adept at capitalizing on emerging market trends. Often this can be accomplished most quickly by leveraging the expertise of third-party providers already on the ground in these new markets (Lee, 2004). The early warning provided by a demand sensing capability allows a supply chain company to make supply chain structural decisions before competitors are fully aware of game-changing trends affecting demand and supply-side decisions.

Demand is estimated in supply chains through forecasts. An unfortunate reality of forecasts is that even forecasts developed from the most sophisticated models using state-of-the-art methods are inaccurate. Forecasts can be improved to better support demand sensing by using *range forecasts*, or forecasts that calculate results for a range of possible outcomes. This technique forces the firm to consider multiple demand possibilities that drive contingency planning across the supply side of the business, including the consideration of options such as flexible contracting, multisourcing, and risk pooling (Sheffi, 2005). Collaborative forecasting options such as CPFR and sales and operations planning (S&OP) provide additional opportunities to improve forecast quality through tighter communication externally (CPFR) and internally (S&OP).

Global Optimization Strategy

Supply chains have lengthened substantially in recent years as supplier, manufacturing, and customer service origination points have moved farther away from consumer points of sale. This movement—characterized by the transfer of much of the manufacturing capacity formerly found in Europe and North America to Asia—has been driven by the search for the lowest cost of production options. Other functions, such as the many customer support and back office support operations that have grown rapidly in India have further expanded the length of supply chain activities. An outgrowth of this approach has been an increase in other costs, especially transportation and inventory, necessary to move products longer distances to market, and store additional product necessary to offset longer replenishment lead times.

As a result, many leading supply chain organizations have begun to question the viability of these lengthened supply chains and are bringing a renewed interest to optimizing the entire global supply chain rather than individual parts of it. A global optimization strategy requires the organization to perform a trade-off analysis to determine the supply chain configuration that provides the best cost-service outcome.

This optimization may take the form of a Total Landed Cost (TLC) comparative analysis. TLC attempts to sum all costs associated with making and delivering products to the point where they produce revenue (Langley, 2010). A comparative analysis strives to compare the total cost of acquiring, transporting, storing, and preparing a product for sale across multiple supply chain configuration options. All costs must be clearly known to produce an accurate analysis. A limitation of a TLC comparative analysis is that although the true costs of the current supply chain may be known, costs of alternative options will have to be estimated, as will future costs. The error associated with these estimates may cloud the analysis; however, an alternative demonstrating a significant savings versus the current option may deserve closer attention.

A related concept is Total Cost of Ownership (TCO). TCO is often applied to supplier-buyer relationships in the supply chain and emphasizes a focus on understanding all the costs associated with acquiring a product (similar to TLC) rather than the traditional price-only focus (Engel, 2011). Partnering should be used to strengthen key supplier relationships in order to refine and streamline processes to ensure that supplier operating costs can be minimized when servicing your firm. Ultimately, the supplier costs become your firm's costs.

ABC Segmentation Analysis

A major business trend emerging out of the last decade is that customer demand has become increasingly fragmented. One-size-fits-all service doesn't meet the needs of today's discerning customer. The most effective supply chain organizations have recognized the trend and have evolved their operations as a result. *ABC segmentation analysis* has been used by many of these organizations to assist in restructuring supply chain operations to fit this new world of demanding customers.

Segmentation implies a method of grouping customers, as described in the previous discussion of the segmentation principle. An ABC segmentation strategy takes a portfolio approach to grouping and managing subsets of customers that have been classified based on a defined set of criteria (Thomas, 2012). The goal of the segmentation is to create groups of customers that have similar value propositions and can therefore be served using similar processes. A typical ABC segmentation may attempt to classify customer sets based on their evaluation in a two-dimensional space, as shown in Figure 3-2. Service requirements, shown on the vertical axis, include an analysis of one or more criteria such as price, sales volume, order characteristics, and delivery requirements.

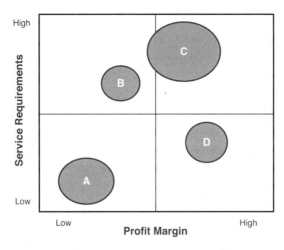

Figure 3-2   Customer segmentation analysis matrix

An effective ABC segmentation strategy is not about the creation of multiple supply chains, each with its own set of distinctive capabilities. Rather, strategic segmentation strives to leverage common infrastructure while building in flexible, configurable elements to serve the unique needs of specific customer groups. When supply chain capabilities are developed in alignment with customer needs, the result should be a high-service, cost-effective supply chain operation.

An example of a company that has benefited from the use of strategic supply chain segmentation is provided by Dell. Over the course of the last several years, the company has transformed its supply chain operations for its former (world class) direct-to-consumer model to a more comprehensive multichannel model with differentiated policies and services designed to serve consumers, corporate customers, distributors, and retails. As a result, Dell has saved $1.5 billion in supply chain costs (Gartner Inc., 2010).

Outsourcing Strategy

Outsourcing has been a significant consideration in supply chain operating strategies for many years. The original rationale driving outsourcing decisions typically included (a) the search for lower-cost manufacturing options, often resulting in the establishment of an operation in China or (b) the desire to establish a presence in emerging markets in order to capture a portion of a new, rapidly growing customer base (Burton, 2013). As markets have continued to evolve the cost-benefit rationale for many of these early decisions has been reversed in the face of rising oil prices and inconsistent product quality. Despite this, the outsourcing of supply chain activities has continued to grow and will be an important component of supply chain strategies into the foreseeable future (Langley, 2013).

Outsourcing allows firms with supply chain operations to focus on their core competencies, which are often not found in the supply chain area, and leverage the logistics and operational expertise of third-party companies (Lieb & Butner, 2007). A well-planned outsourcing strategy may provide many benefits to the firm including the following:

- Rapid entry into new markets
- Rapid changeover to new service offerings
- Reduction in capital required to expand or alter supply chain operations
- Reduced cost of materials and operations
- Access to improved IT and data-analysis capabilities

Historically, many outsourcing decisions have been made piecemeal with a short-term, tactical mindset. Over time, the result of multiple independent decisions to outsource functions or add operations in new markets has led some global companies to discover that they were coordinating third-party contracts with dozens of organizations (Langley, 2010). A strategic view of outsourcing requires the organization to make a decision on what scope of outsourcing it desires and which functions and geographies should be considered. The scope decision spans a wide spectrum from outsourcing a single function, to outsourcing multiple functions using a single provider, to outsourcing multiple functions using multiple providers. The appropriate option selected is completely dependent on an accurate assessment of the organization's internal supply chain capabilities and strategies that need to be supported.

Apple is an example of a company that has utilized outsourcing effectively to create a world-class supply chain. Apple sources raw materials and manufactures the majority of its products in third-party facilities in Asia and uses third-party transportation carriers to move finished goods to retailers and end consumers. Apple has frequently been mentioned as one of the top supply chain companies in the world.

Mass Customization Strategy

Mass customization provides a manufacturing strategy that blends the economies of scale found in mass production with the product uniqueness many customers prefer. A mass customization strategy seeks to produce a variety of product variants at, or close to, the cost associated with less flexible traditional production methods. It is defined as the capability to manufacture a relatively high volume of product options for a relatively large market that demands customization, without trade-offs in cost, delivery, and quality (McCarthy, 2004).

Several approaches are possible for firms wanting to pursue a mass customization strategy. Individual customization of products is the limit providing for complete customization of each product produced. An example of individual customization is found in

Dell's fulfillment of laptop orders. Customers "design" their laptop using the Dell website ordering system online by choosing from a range of options for each component that will ultimately go into the device. Once the product is completely configured in the order environment, it is sent to a Dell assembly facility and batched with similar orders. Customers must accept two minor limitations. First, production of the product will typically be delayed a few days until enough orders with similar configurations are accepted. Second, the customer is not provided unlimited options in configuring the device. If the company attempted to relax either of these constraints, production costs would no doubt skyrocket.

At the other end of the mass customization continuum are companies that choose to product and sell a large variety of standard products. In this case, postponement is an often-used complementary technique that allows final product configuration to be delayed until the product is moved closer to the point of customer demand. In both build-to-order and configure-to-order scenarios, companies that choose to mass customize generally opt to simplify their upstream supply chain by decreasing the variety of parts used in the production process.

Technology Integration

The ability to make strategic decisions quickly based on accurate, up-to-date information is a critical aspect of supply chain management (Mentzer, 2004). Information technology consists of the tools used to gain awareness of information, analyze the information, and execute on it to increase the performance of the supply chain (Chopra & Meindl, 2013). Information is the lifeblood of the supply chain and is needed to drive virtually every process. Without technology and information-generating resources, supply chain managers would lack the visibility required to manage the environment.

Technology provides many benefits, but three stand out as vital. First, technology and information facilitate collaboration and tighter integration between supply chain partners. This is because information sharing provides for greater transparency in supply chain relationships, in addition to supporting the creation of more consistent, streamlined processes. Technology-enabled information sharing allows managers to focus on ways to improve the relationship.

Second, it provides visibility upstream and downstream, including visibility to customer demand, forecasts, in-transit inventories, and supply delays. Without this information, each organization in the supply chain would be much more subject to the bullwhip effect, allowing minor ripples to become major disruptions.

Third, technology provides processing speed that allows enormous volumes to be handled. Bar codes, RFID, and customer-initiated order entry via the Internet are examples of technology that has greatly reduced or eliminated manual data management activities, allowing items to be handled and moved more quickly.

The importance of technology will continue to grow as new technologies are developed and adapted for use across the supply chain. The use of the Internet in B2B collaborations is an example of a technology that has been leveraged to improve interactions between organizations (Muckstadt, Murray, Rappold, & Collins, 2003). The technology that will make the next big impact is impossible to guess, but you can be sure that one or more new technologies will arrive on the scene and change the way supply chain organizations do business in the coming years.

Channel-Spanning Performance

Performance is defined as the evaluation of how previously established goals have been met (Mentzer & Konrad, 1991). Certainly, companies establish goals that in turn are used as a measuring stick with which to gauge performance. Goal creation and subsequent performance measurement are a bit trickier when viewed from a supply chain perspective because these goals need to be mutually agreed upon by supply chain partners, and performance evaluation should include metrics that span the entire supply chain (Defee & Stank, 2005). A channel-spanning performance strategy requires the creation and monitoring of KPIs intended to evaluate performance across the whole supply chain.

Determining macro performance of the supply chain is a daunting task to some, but several techniques exist that can provide the information and insight needed. The effectiveness of the supply chain can be seen as the ability to produce quality customer service, and this can be evaluated using the perfect order metric (Anderson, Britt, and Favre 1997). Perfect order is simply an order that arrives at the customer location on time, complete, undamaged, and priced correctly. In effect, the perfect order metric spans all activities across the entire supply chain because ultimately the purpose of the supply chain is to satisfy and serve consumer demand, and this metric measures how effective the cumulative efforts of the supply chain have been at meeting this goal. Perfect order is also an example of a downstream or customer-facing metric (Hoffman, 2007).

Upstream or supplier-facing metrics are also needed. Supplier scorecards provide logically organized, consistent, composite reporting of performance. A scorecard facilitates the comparison of KPI results across multiple suppliers and can serve as an early warning system to help pinpoint the location of problems. The process of scorecarding can improve the relationship with suppliers when used as a mutual problem resolution tool.

Benchmarking performance against an industry database can provide real insight into how the company and its end-to-end supply chain compare with competitors (Hoffman & Barrett, 2010). Of critical importance is determining the distinctive characteristics of the company's supply chain and making a benchmark comparison against supply chains with similar structure. Routine benchmarking may be the best process for identifying positive versus negative trends of the supply chain enterprise.

Complexity Reduction

The supply chain environment is a complex one. A host of factors has added to the complexity supply chain that executives must deal with, including the growth of outsourcing, lengthening lead times caused by increasingly distant supplier and manufacturing locations, more frequent new product introductions, fragmented demand, and fuel price volatility, to name a few. To compound this, large organizations, especially retailers, must manage hundreds or even thousands of supply chains. Unless the growing complexity is managed proactively, it can become "the cancer that destroys supply chain efficiency" (Gilmore, 2008).

A complexity reduction strategy is a necessary yet practical tool available to supply chain executives to simplify the supply chain environment. A variety of simplification options have been proposed such as designing products to use common component parts, reducing product variety, and reducing demand forecast error. Two powerful options available to complexity managers are supplier consolidation and nearshoring.

Supplier consolidation reduces the number of organizations that a supply chain company must interact with to source and move its products. The simplified supply chain environment created through a pairing of the supplier base has produced benefits including reduction in transaction costs, lessening of risk, more consistent product quality, and increased supplier innovation and responsiveness (Choi & Krause, 2006). Consolidation activities should spread beyond suppliers of raw materials and finished products to include all entities in the supply chain, including third-party service providers, transportation carriers, and distributors.

Rising crude oil prices and the associated increase in transportation costs during the past decade have shifted the economic basis for many organizations that chose to move manufacturing operations to low-cost areas like China. *Nearshoring*, or the location of supply and manufacturing closer to the points of demand being serviced, has become a viable option for many organizations as a result of these shifting economics (Simchi-Levi et al., 2012). Locating upstream operations closer to demand shortens lead times, reducing inventory requirements and producing a more responsive supply chain.

Cross-Chain Collaboration

Many leading supply chain organizations have sought to break the long-standing adversarial paradigm that has existed for decades between buyers and seller by developing closer relationships with key supply chain partners. Collaboration among supply chain organizations has led to significant reductions in inventory, increased response time, and improved service levels. *Collaboration* is defined as the joint initiatives that go beyond the normal course of day-to-day business, with the aim of delivering significant improvements over the long run (Benavides, De Eskinazis, & Swan, 2012). Effective collaborations between firms lead to tighter integration and improved customer service performance (Stank, Keller, & Daugherty, 2001).

Cross-chain collaborative efforts may create a unique capability that provides a competitive advantage for the partners involved (Fawcett, Fawcett, Watson, & Magnan, 2012). Collaborations may involve buyer-seller (for example, supplier-manufacturer), also called vertical collaborations, or associations among organizations occupying the same echelon (for example, two or more manufacturers working together), also called horizontal collaborations. A variety of barriers must be overcome for a collaboration to be successful, including an unwillingness to share information, lack of trust, and a perceived unfairness in effort or benefits accruing to the partners.

An element of successful strategic collaborations that is often overlooked is the importance of selecting partners that are seeking to establish strategic alignment with your firm. Successful collaborations require that mutually held goals be established between the partners. Without this common focus on goal achievement, collaborations will fall short of their potential. Implementing a successful collaboration first requires the "right" partner(s) with appropriate skillsets be identified (Benavides et al., 2012). The largest potential partner may not be the best fit given the capabilities, goals, and interest of both companies. Successful implementation also requires finding the "right" opportunity. One that provides real benefit to all parties and can lay the groundwork for future joint efforts has much greater potential than trying to work together to eliminate the pain of a one-off problem area.

Lean Logistics

A lean philosophy targets the elimination of waste in the supply chain through the creation of smoother flowing, faster, more efficient processes. In a supply chain, this requires commitment of multiple companies working together to implement the philosophy. *Lean logistics* focuses on the movement and storage aspects of supply chain activities. Lean logistics is defined as a strategy by which companies seek to create a level flow of material through the supply chain based on smaller, more frequent shipments, in support of one of the central tenants of lean manufacturing: carrying as little inventory as possible (Gilligan, 2004).

Multiple lean concepts can be applied to logistics activities to create a lean logistics strategy and reduce inventory throughout the supply chain (Srinivasan, 2004). The most relevant concepts and their application to lean logistics are these:

- **Produce only what is needed (*Kan-ban*)**—Create a pull system by working to demand triggers rather than forecasts. Share demand information across the supply chain.

- **Eliminate waste (*muda*)**—Inventory is considered waste in a lean logistics context. Reducing cycle times and minimizing storage points supports this principle.
- **Smooth the flow of work (*heijunka*)**—Level out activity by eliminating spikes in demand.
- **Mistake proofing (*poka yoke*)**—Address ineffective processes and improve integration though the introduction of automation.
- **Continuous improvement (*kaizen*)**—Focus on making small, routine improvements in processes.

Inventory accumulates at points in the logistics network where it stops moving toward the ultimate consumer, generally in warehouses and distribution centers. Although warehouses serve a number of functions, a major reason for the existence of warehouses and DCs is because inventories stored there serve as a buffer against unanticipated demand. Lean logistics seeks to create a demand-driven environment and keep inventory moving and thus eliminate the need for traditional warehouses where possible. Flow-through facilities and cross-docks provide for sortation and aggregation while allowing product to move quickly through nodes in the supply chain.

Risk Management and Contingency Planning

Service disruptions are likely to occur, but by definition the timing and scope of disruptions cannot be anticipated. This is not to imply that decision makers are not prepared to deal with the risk(s) presented by a disruption. Leading supply chain organizations utilize contingency planning to avoid being caught off guard when unplanned events happen.

Not surprisingly the list of *leading supply chain companies* that develop contingency strategies for supply chain operations is fairly short. In a recent study (Cudahy, George, Godfrey, & Rollman, 2012), only 11 percent of companies surveyed reported that they actively manage supply chain risk. Only 18 percent noted the existence of a formal supply chain risk management system.

The starting point for contingency planning is to understand the range of possible risks of disruption the supply chain may face in the future. These risks can be categorized based on the degree to which they can be anticipated and whether the organization can influence the occurrence of the potential disruption, as shown in Figure 3-3. This can be approached as a white board exercise. Additional potential risks can be added over time as they are identified. The next step is to identify the vulnerabilities of the supply, particularly as they relate to the identified risks. Each supply chain has vulnerabilities and it's imperative that decision makers provide an honest assessment of these so that plans to address them can be properly prioritized.

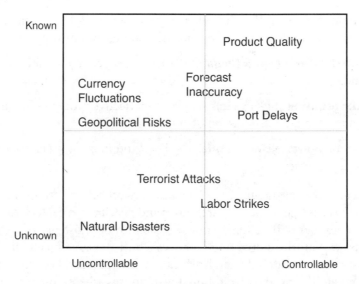

Figure 3-3 Supply chain risk assessment

As noted in the results of the Accenture research mentioned previously, few firms take the next step of actually defining and documenting contingency plans that link to each of the identified risk areas. Each contingency plan should lay out the immediate actions to be taken when the disruption being addressed occurs. It should provide a road map of what decisions need to be made and in what order. The plans should be reviewed by key decision makers, and any missing details should be added. Finally, the plans should be shared with managers throughout the organization and those from key partner organizations that will have to execute the plans in the future. All plans should be reviewed and updated on a routine cycle of 1–2 years.

Supplier Sustainability Audits

Supplier compliance programs have evolved over the years. Originally, most compliance programs were simply punishment tools used to assess penalties on suppliers that provided less-than-desirable service quality or were not fully compliant with the company's rules. The punishment focus is out and reinforcement is now in vogue as supplier compliance now tends to emphasize and reward doing things right rather than stressing the wrong. These programs are once again changing as many large companies have begun to emphasize the sustainability. For the most part, service quality is expected, and supplier audits are focusing on performance of broader business strategies such as sustainability.

Sustainability is broadly defined as development that meets the needs of the present without compromising the ability of future generations to meet their needs (World Commission on Environment and Development, 1987). Company policies of sustainability likewise cover a wide area and include social stewardship elements such as environmental, social, health and safety, ethical, and economic (cost-effective) performance. Major companies have become the drivers of sustainability throughout the business world. Wal-Mart announced its plan to emphasize sustainability in 2005 and has since announced more than 35 specific sustainability targets it hopes to meet (Fishman, 2007). The majority of these sustainability targets require the company to influence the behavior of its 100,000-plus suppliers and third-party service providers. Sustainability audits ensure that suppliers comply with Wal-Mart sustainability programs.

Barriers to Success

A *barrier* is defined in many ways: a boundary or limitation; something that obstructs progress or access; something that separates or hinders union; or, any condition that makes it difficult to make progress or to achieve an objective. These impediments to success can exist naturally, be manmade, or be a combination of both (TheFreeDictionary, 2013). They can also be created within or outside the organization. Internal barriers are largely caused by fear and protectionism, poor information flows, misalignment of strategies and rewards, short-term thinking, and rogue financial objectives (Rieger, 2011).

Regardless of their sources and causes, barriers make it difficult for the organization to successfully execute strategies and achieve desired goals of efficiency, responsiveness, and growth. The organization must prioritize efforts to break free of these impediments and make progress toward strategic goals. This can be done by conducting an objective analysis of the root cause, manifestation, and impact of barriers. Then, the barriers must be prioritized for removal based on their influence and how difficult they will be to eliminate (Rieger, 2011).

It is absolutely critical to reduce supply chain barriers to success, according to a recent report. The report states that supply chain barriers are more significant impediments to trade than import tariffs. Supply chain barriers result from inefficient customs and administrative procedures, complex regulation, and weaknesses in infrastructure services, among many others. Lowering these barriers eliminates resource waste, reduces costs, and creates lower consumer prices. Global GDP would increase 4.7 percent if supply chain barriers were cut in half (World Economic Forum, 2013).

In this section, 10 harmful barriers to supply chain management success are analyzed, with a specific focus on their descriptions, causes, and consequences. Initiatives to remove these barriers are needed so that critical supply chain strategies can be implemented and processes can be executed as intended. An inability to eradicate, reduce, or work around performance-sapping barriers like these 10 issues will make it extremely difficult for an organization to maximize supply chain success.

Network Complexity

One of the fundamental challenges of supply chain management is the growing complexity of supply chain networks. As Chapter 1 and Chapter 2 have revealed, a supply chain involves numerous organizations and multiple processes, each with its own priorities and perspectives on best practices. As the scope and scale of a supply chain expands with additional suppliers, products, customers, and locations to serve, the overall complexity and risk levels quickly rise.

The primary contributors to this supply chain barrier are the business trends of globalization, outsourcing, and consumer dynamics. Offshore manufacturing and global market expansion add distance, time, and new requirements to a supply chain. The use of contract manufacturers and logistics service providers increases the fragmentation of ownership and boosts the number of stakeholders in the system (Chopra & Meindl, 2013). Consumers' omnichannel shopping habits, demand for greater product variety, and desire for product customization each require modification or expansion of supply chain production and fulfillment capabilities. These three trends can lead to supply chain network expansion and operational scope creep that makes the system more convoluted.

The outcomes of a highly complex supply chain tend to be negative. First, the increased distance and number of facilities impedes supply chain visibility. It becomes more difficult to maintain control of the network, and the risk of service disruptions rises. Second, the increased number of participants increases communication challenges. A lack of effective communication makes it harder to align processes along the supply chain and increases the risk of the organization becoming more insular and focused on its own interests to the detriment of overall supply chain performance. Finally, an expanded set of fulfillment channels and capabilities is costly to develop and difficult to manage. Organizations must vigilantly work toward streamlining their supply chains to avoid the systems complexity barrier and its associated consequences.

External Pressures

As supply chain management has gained greater attention inside the organization, so too has the level of scrutiny from the media, government agencies, and other parties that previously showed little interest in this field. These outsiders are putting great pressure on organizations to consider the impact of supply chains beyond the corporate income statement. A failure to respond in a timely and constructive manner can lead to negative press, governmental sanctions, and customer defection.

The drivers of this external pressure include state interests, expansive sustainability campaigns, and a growing localism philosophy. First, governments are putting more regulatory emphasis on supply chain activity. Increasing transportation safety regulation, environmental defense directives, consumer protection laws, and trade barriers affect sourcing decisions, product flows, total costs, and other supply chain considerations. New sustainability initiatives, often led by nongovernmental organizations, go well beyond environmental considerations. For example, a low-cost country sourcing plan must also consider anticorruption, human rights, and traceability issues. Finally, the growing call for locally produced goods and locally owned, small businesses is pressuring organizations to respond in a positive manner. Wal-Mart, for example, has embarked on a supply chain initiative to double the amount of fresh produce from local growers by 2015 (Reuters, 2013).

The primary effect of these external pressures is a need to think beyond the interests of core supply chain stakeholders. No longer can supply chain managers base their decisions solely on corporate interests. Regulatory compliance, societal impacts, and reputational considerations will need to figure more prominently in supply chain analysis, design, and operation. This may, of course, slow the planning process, drive modification of existing methods, and add cost to the system. However, if external pressures are successfully handled with a balance of logic and sensitivity, an organization's supply chain will be viewed as a solution rather than an impediment to the advancement of key societal goals. As a result, the former barrier can become a building block of competitive advantage for the organization.

Talent Gaps

Supply chain strategy tends to emphasize processes, product flows, and information transfer, but the common lynchpin of success is the people who make the strategic decisions and execute them. As the supply chain grows in importance to top management, a shortage of broadly skilled supply chain talent becomes a major impediment to success. Though most organizations have people with strong fundamental skills in supply chain operations, gaps in big-picture thinking, analytics, and resourcefulness will make it extremely difficult to manage global network complexity, coordinate omnichannel processes, and engage external stakeholders (Gibson, Williams, Goffnett, & Cook, 2013b). Stronger interpersonal skills are also needed to overcome the people issues—culture, trust, aversion to change, and willingness to collaborate—that hamper supply chain initiatives (Fawcett, Magnan & McCarter, 2008).

Talent gaps arise for many reasons. One major cause is the lack of focus on supply chain talent by top management. Historically, supply chain managers were converts from other disciplines and had limited formal training. Prioritization of investment in supply chain technology, information, and measurement systems over management development also contributes to the talent gap. The third contributor is the misperception that talent development is the domain of human resources and the individual. This leads managers to abscond their duty to build a strong pool of promotable supply chain professionals. When supply chain talent perceives that the organization is not investing in their capabilities, they seek out better opportunities elsewhere and the internal skill set is further depleted.

The immediate impact of a talent gap is an inability to develop strong internal supply chain capabilities. Senior leaders can make great promises to C-level executives, but without talented people to create and execute the plan, little will be accomplished. A lack of talent can also lead to an overreliance on other organizations across the supply chain. Their strategies and processes will dictate the capabilities of the supply chain and may lead to higher overall cost for the organization. The net results of weak development programs are supply chain talent turnover and an even wider talent gap. Talent flight will include the most highly skilled individuals, weakening the organization's pool of future supply chain leaders (Gibson et al., 2013a).

Process Disconnects

To capture the cost and service benefits of supply chain management, the processes identified in Chapter 2 must be tightly coordinated. Activities must be linked, demand signals and other data must be shared, and functional strategies must be compatible. In contrast, when little effort is made to connect supply chain processes within and across organizations, alignment is difficult to achieve. Autonomous process development often produces incompatible goals, misaligned schedules, and operational bottlenecks. Most important, a disconnect between supply and demand emerges that hampers the ability of the supply chain to achieve the fundamental goals outlined in Chapter 1: efficient fulfillment, customer value, organizational responsiveness, and network resiliency.

Tradition and fear play major roles in disconnected supply chains. The field evolved from a mixture of internal functions that operated independently and sometimes competed for resources. Add in a variety of external participants that may not be willing to share information, connect systems, or develop compatible procedures, and the makings of a nonintegrated and potentially dysfunctional supply chain emerge. Ultimately, connecting the people, processes, and technologies across a supply chain is both time-consuming and expensive. Unless a strong supply chain leader can overcome these concerns with a strong vision for process alignment, it will be difficult to gain buy-in and action from affected parties.

Supply chain process disconnects generate numerous consequences and related financial repercussions for an organization. When a supply chain is not synchronized around demand response, production managers base their activities on forecasts and internal production schedules. Ineffective forecasting can lead to incorrect allocation of limited production capacity, excess finished goods inventory, and a backlog of customer orders that require expediting. In turn, higher supply chain costs are incurred. The inability to respond rapidly to changing market conditions is another flaw of disconnected processes. For example, if the fulfillment and transportation capabilities are not agile enough to rapidly shift inventory between regions, the organization will miss out on revenue-generating opportunities. In both cases, the income statement is negatively affected by the incompatible, ineffectively coordinated supply chain.

Technology Deficiencies

Though it is typically considered to be an essential supply chain tool, technology is a double-edged sword for managers. On one hand, an effective combination of computing power and supply chain software gives organizations the ability to pursue optimal network design and performance. On the other hand, it is a great challenge to purchase and implement systems that can cost-efficiently support current supply chains and adapt to future requirements. In the current environment, outdated systems cannot effectively support employee use of mobile computing tools, omnichannel fulfillment of customer demand, or in-depth analysis of the growing cache of supply chain data.

An organization's inability to harness the benefits of technology is usually the result of internal failures. Some organizations are overly reliant on proprietary supply chain systems that are difficult to upgrade, while others use outdated planning and execution software that is not readily compatible with the organization's Enterprise Resource Planning (ERP) system. Both issues make it difficult to connect and collaborate with supply chain stakeholders. Another cause of supply chain technology strife is the low priority placed upon maintaining and improving these systems by the information technology group. Finally, a lack of spending on upgrades and new systems increases the velocity with which systems lose their effectiveness.

The supply chain repercussions of poor technology capabilities are significant. A lack of cross-chain connectivity will hamper visibility of sales, inventory levels, and product flows. This limits the organization's ability to sense and respond to demand with inventory that is available across the supply chain. Also, outdated tools do not offer event management capabilities that are needed to quickly alert managers to problems and help them alleviate service disruptions. Another potential problem is the inability to accurately model existing processes that render supply chain planning difficult, slow, and ineffective. Ultimately, an unwillingness or inability to pursue advanced supply chain systems will make it impossible for an organization to fully exploit the technology capabilities and innovations discussed in Chapter 4.

Transactional Focus

Cross-chain collaboration is a widely touted strategy in supply chain management because it facilitates process integration, efficiency, and a common focus. The problem is that many organizations continue to operate in a highly transactional mode in which engagements with suppliers and customers are treated as one-time events. These transactions are price-driven, little effort it made to coordinate activities, and a compliance focus emerges. As the transactional focus becomes institutionalized, suppliers and customers develop little loyalty to the organization, and a major barrier to supply chain improvement emerges.

The transactional focus often results from an adversarial approach to procurement and distribution. When an organization views every transaction as a win-lose opportunity, and managers are incentivized to squeeze suppliers, transportation providers, and customers for better deals, there is little chance of a relationship-driven approach to external stakeholder alignment emerging. Another contributor to the transaction focus is a lack of history between selling and buying organizations. Sellers must deliver on commitments over time to create a level of trust before the buyer will consider moving beyond a transactional focus. Finally, there may be a size discrepancy between the organizations that makes anything more than a transactional focus impossible. When a major retailer such as Tesco has thousands of suppliers, it is not feasible to set up collaborative relationships and customized supply chain processes with its small volume vendors. Instead, Tesco must limit its strategic collaboration to its largest and most impactful suppliers.

The result of a transactional focus is a blockade to strong working relationships where information sharing and chain-wide knowledge drive action. Cross-chain collaboration does not occur where trust does not exist, commitments are not formalized through contracts, and a win-win perspective is adopted. Without these relationship elements, there will be little motivation to pursue chain-wide optimization efforts or establish technology links between the organizations. As a result, many of the key supply chain principles discussed earlier—visibility, agility, alignment, integration, and so on—will not be achieved to a meaningful level.

Disparate Goals

To be successful, a sports team needs to have a common vision of winning that drives the efforts of each player. When players focus instead on individual accomplishments and statistics, their actions become self-serving and potentially destructive to the overall goal. The same problem occurs both inside and outside the organization. Inside the organization, managers of different processes may prioritize goals that optimize functional performance rather than supply chain performance. Externally, each organization tends to set goals independently and with a focus on its own financial results. This makes it particularly difficult to convince top executives to invest in the supply chain, take on a larger role, or incur short-term expenses to improve its long-term efficiency.

The fundamental driver of disparate goals is poorly aligned incentives and metrics. When performance evaluations, bonuses, and career interests are at stake, it is human nature to focus on the goals that incentivize individual interests whether or not they contribute to overall supply chain performance. For example, production managers are rewarded for labor productivity and maximum output, which drives them to pursue long production runs of similar products. This is in conflict with an inventory manager's charge to match supply and demand with little safety stock. Another contributor to disparate goals is poor communication. Unless there is a supply chain champion who can clearly articulate a common vision for the supply chain and its value to all stakeholders, there is limited chance to break free of disparate goals and independent incentives. Finally, inconsistent priorities and definitions of success can lead to service quality and cost efficiency goal conflict.

The consequences of disparate goals affect stakeholders across the supply chain. Internally, functional silos emerge and performance targets revolve around them. The lack of cohesive internal goals around total cost, revenue growth, and profitability will impede supply chain performance. Externally, an inability to align stakeholders with a consistent supply chain vision and goals will make it difficult to synchronize information and product flows. Each organization tends to its own shareholders and pursues profitability targets without the benefit of an integrated supply chain. The results will be less than optimal and supply chain innovation will be stunted.

Functional Silos

In an organization suffering from a *functional silo syndrome*, the interactions of each department or process are focused inwardly. Communications, strategies, and decisions are made inside the silo with little effort put forth to engage other groups across the organization. This lack of engagement breeds insular thinking, redundancy, suboptimal decision making, and subculture development (Rosen, 2010). The disparate goals discussed earlier also tend to emerge. These functional silos are particularly problematic from a supply chain standpoint because the principles of integration and efficiency dictate that the plan, buy, make, fulfill, and return processes should be aligned and avoid duplication of effort. When silo barriers are in place, a supply chain does not truly exist and it is difficult to effectively serve customers.

Silos are organization-induced problems caused by traditional structural designs and restrictive policies. Setting up an organization with functional processes, goals, and reporting structures hinders collaboration across areas. Also, separating functions in different facilities and geographic locations, creating independent supply chains for each operating division, and promoting competition between areas tend to reinforce the silos. Finally, formal policies fuel silos and poisons collaboration because people become overly concerned with protocol and politics. For example, when policies discourage individual supply chain managers from engaging their counterparts in marketing, sales, manufacturing, and related areas without first going through leadership channels, unnecessary barriers are created and isolation is perpetuated.

A damaging consequence of a silo mentality is the loss of cross-functional engagement and innovation. A lack of engagement can generate internal competition between silos that squelches collaboration, information sharing, and trust. Silos can also extend to a function's information systems and data. When critical information about demand, inventory, schedules, capacity, and performance is not shared among supply chain functions, the data becomes trapped in and unavailable to managers. This leads to unnecessary guesswork and fractured decision making. Ultimately, these functional silos can cost an organization in lost supply chain agility, productivity, and responsiveness.

Bullwhip Effect

The *bullwhip effect* is a major barrier to supply chain success that feeds upon many of the previously discussed impediments. The bullwhip effect occurs when demand variability is amplified across the channel due to inaccurate demand sensing and ordering by upstream suppliers. Small changes in consumer demand and retailer orders set off a series of subsequent orders where the ordering pattern of each upstream supplier is greater than that of its downstream customer (Lee, Panmanabhan, & Whang, 1997). Though the problem is widely recognized, organizations continue to distort upstream demand signals and deviate from logical demand driven ordering processes which leads to tremendous inefficiencies across the supply chain.

The problem is caused by the infrastructure of the supply chain, the order managers' rational decision making, and four primary drivers. First, the inclination of each organization is to independently forecast demand based on its customer's order and its safety stock requirements. This forecast magnifies the forecast errors built into downstream orders. Second, batching orders creates periodic surges in demand that are difficult to interpret and amplifies the variability problem. Third, price fluctuations such as discounting encourage ordering patterns that do not match consumption patterns. The variation of the buying quantities is much bigger than the variation of the consumption rate, magnifying the bullwhip effect. Finally, when product is in short supply, customers may place orders that far exceed real demand to ensure inventory availability. These inflated demand signals set off upstream production and procurement activities that are not truly needed (Lee et al., 1997). Other causes include multiple echelons in the supply chain and long lead times.

When a supply chain experiences the bullwhip effect, the outcomes are costly. The distorted and incomplete demand information often causes organizations to overreact in their inventory planning. Overordering leads to excess inventory in the network and higher inventory carrying costs. Demand drops cause order cuts and cancellations that leave store shelves understocked. In addition to the inventory headaches, these boom-and-bust ordering tactics create tremendous inefficiencies across supply chain processes: an inability to accurately determine capacity needs and meet production schedules, the need for expedited transportation, and poor customer service. The ultimate impact is higher cost and lost sales.

Limited Visibility

Visibility is an important supply chain principle that is cited by managers as essential to reducing costs and improving operational performance. It's crucial for the tight synchronization of supply and demand, including the movement of goods, information, and funds (Miles, 2013). Despite the interest, potential benefits, and variety of technology options, many organizations struggle with this principle and suffer from limited cross-chain visibility. This supply chain barrier makes it difficult for managers to monitor and control events outside their individual operations. After an order has been placed or inventory has been given to a transportation company, managers in a low visibility supply chain encounter many blind spots and must patiently wait for information. They can only hope that key processes are being properly executed by other organizations.

These visibility challenges are created by technology and data issues. Though visibility tools can be helpful, some systems are challenging to implement and costly to integrate with external stakeholder systems. Also, vendor promises may exceed their current capabilities. For example, they may offer only separate, fixed views of information that cannot be integrated for a complete picture of the supply chain. Others tout their sense-and-respond capabilities but provide only a limited level of visibility and depend on the users to determine appropriate responses. When the systems do work well, information overload issues can overwhelm organizations that are unprepared for the daily data avalanche. Front-line staff must be able to easily consolidate, modify, and analyze visibility information to quickly respond to change (Kinaxis, 2011).

Poor supply chain visibility has negative repercussions for an organization that is trying to satisfy customers in highly competitive markets. When supply chain managers cannot effectively track in-transit freight, they get no early warning signals of disruptions and are forced into a reactive response mode. Inadequate traceability methods can become a compliance problem as government agencies are ramping up visibility requirements for at-risk products such as pharmaceuticals, meat, and fresh produce. Poor visibility also impedes an organization's capability to analyze cross-chain cycle time and costs for improvement initiatives. Finally, limited visibility of demand contributes to the bullwhip effect as managers lack the necessary data for forecasting and ordering purposes.

Chapter Summary

Managing a supply chain is not an exercise in optimizing the individual plan-buy-make-move-return processes of suppliers, manufacturers, and customers. Instead, savvy managers focus holistically on the supply chain as a connected entity and pursue tangible outcomes based on value creation for the customer and profitable growth for each link in the supply chain (Anderson, et al., 1997). This enduring concept is foundation for the supply chain principles and strategies discussed in this chapter.

The journey toward a profitable, value-creating supply chain begins with a set of widely accepted principles. These fundamental truths provide an understanding of how highly effective supply chains are structured, operated, and improved. Twelve important supply chain principles, ranging from agility to sustainability, were highlighted in the chapter. These principles help managers concentrate on the capabilities and critical success factors that drive decision making and promote long-term viability of the supply chain.

Next, managers must translate the principles into logical supply chain strategies that align with goals of the organization, support the needs of customers, and account for the challenges posed by competitors. These strategies must provide a path forward by establishing a set of actions that create the supply chain capabilities needed by the organization in the future (Dittmann, 2012). A dozen well-tested strategies for creating supply chain value were discussed in the chapter. They cover a wide range of planning and execution issues capable of propelling the supply chain toward its stated objectives.

When investigating principles and developing strategies, supply chain managers must be aware of the potential barriers to success. These impediments often emanate from internal people, process, technology, and policy flaws that constrain supply chain capabilities. External alignment and connectivity weaknesses also limit supply chain capabilities. This chapter addressed both internal and external barriers that managers must abate for the good of the supply chain. A failure to do so will perpetuate problems and prevent effective execution of supply chain strategies.

References

Amazon.com. (2013) *About Amazon.* Retrieved September 20, 2012, from http://www.amazon.com/b?node=239364011.

Anderson, D. L., Britt, F. F., and Favre, D. J. (1997) The 7 principles of supply chain management. *Supply Chain Management Review.* Retrieved September 12, 2013, from http://www.supplychainventure.com/old-site/pdf/TheSevenPrinciplesofSupplyChainManagement.pdf.

Benavides, L., De Eskinazis, V., and Swan, D. (2012) Six steps to successful supply chain collaboration. *Supply Chain Quarterly,* 6(2). Retrieved September 12, 2013, from http://www.supplychainquarterly.com/topics/Strategy/20120622-six-steps-to-successful-supply-chain-collaboration/.

Boone, C. A., Craighead, C. W., and Hanna, J. B. (2007) Postponement: An evolving supply chain concept. *International Journal of Physical Distribution and Logistics Management,* 37(8), 594–611.

Boone, C. A., Craighead, C. W., Hanna, J. B., and Nair, A. (2013) Implementation of a system approach for enhanced supply chain continuity and resiliency: A longitudinal study. *Journal of Business Logistics*, 34(3), 222–235.

Burton, T. T. (2013) The 10 hidden costs of outsourcing. *Supply Chain Quarterly*, 7(2), 48–54.

Cecere, L. (2013) A practitioner's guide to demand planning, *Supply Chain Management Review*, 14(2), 40–46.

Choi, T. Y., and Krause, D. R. (2006) The supply base and its complexity: Implications for transaction costs, risks, responsiveness, and innovation, *Journal of Operations Management*, 24(5), 637–659.

Chopra, S., and Meindl, P. (2013) *Supply chain management: Strategy, planning, and operation*. Upper Saddle River, NJ: Pearson Education.

Coyle, J. J., Langley, C. J., Gibson, B. J., and Novack, R. A. (2013) *Supply chain management: A logistics perspective* (9th ed). Stamford, CT, South-Western Cengage Learning.

Cudahy, G. C., George, M. O., Godfrey, G. R., and Rollman, M. J. (2012) Preparing for the unpredictable. *Outlook*, Accenture.

Dapiran, P. (1992) Benetton: Global logistics in action. *International Journal of Physical Distribution & Logistics*, 22(6), 1–5.

Defee, C. C., and Stank, T. P. (2005) Applying the strategy-structure-performance paradigm to the supply chain environment. *International Journal of Logistics Management*, 16(1), 28–50.

Dittmann, J. P. (2012) Developing a supply chain strategy for the years ahead. *Supply Chain Management Review*. Retrieved September 13, 2013, from http://www.scmr.com/article/developing_a_supply_chain_strategy_for_the_years_ahead/.

Engel, B. (2011) 10 best practices you should be doing now. *Supply Chain Quarterly*, 5(1). Retrieved September 13, 2013, from http://www.supplychainquarterly.com/topics/Procurement/scq201101bestpractices/.

Fawcett, S. E., Fawcett, A. M., Watson, B. J., and Magnan, G. (2012) Peeking inside the black box: Toward an understanding of supply chain collaboration dynamics. *Journal of Supply Chain Management*, 48, 44–72.

Fawcett, S. E., Magnan, G. M., and McCarter, M. W. (2008) Benefits, barriers, and bridges to effective supply chain management. *Supply Chain Management: An International Journal*, 13(1), 35–48.

Feitzinger, E., and Lee, H. L. (1997) Mass customization at Hewlett Packard: the power of postponement. *Harvard Business Review*, 75(1), 116–121.

Fishman, A. (2007) Working with the enemy. *Fast Company*, 118, 74–81.

Gartner Inc. (2010, November) Case study for supply chain leaders: Transformative journey through supply chain segmentation.

Gibson B., Williams, Z., Goffnett, S., and Cook, R. (2013a) *SCM talent development: The advance process*. Oak Brook, IL: Council of Supply Chain Management Professionals.

Gibson B., Williams, Z., Goffnett, S., and Cook, R. (2013b) *SCM Talent Development: The acquire process*. Oak Brook, IL: Council of Supply Chain Management Professionals.

Gilligan, E. (2004) Lean logistics: Not a fad diet. *Journal of Commerce*, 18–20.

Gilmore, D. (2008) Supply chain complexity crisis. *Supply Chain Digest*. Retrieved September 13, 2013, from http://www.scdigest.com/assets/FirstThoughts/08-06-12.php.

Hoffman, D. (2007) Supply chain measurement: Turning data into action. *Supply Chain Management Review*, 11(6), 20–26.

Hoffman, D., and Barrett, J. (2010) Benchmarking your supply chain. *Journal of Commerce*, 11(35), 30–33.

Kinaxis, Inc. (2011) *Achieving supply chain visibility: More than meets the eye*. Retrieved September 18, 2013, from http://www.kinaxis.com/downloads/register/WP_AchievingVisibility.pdf.

Kucera, D. (2013) Amazon ramps up $13.9 billion warehouse building spree. *Bloomberg Businessweek*. Retrieved September 20, 2013, from http://www.businessweek.com/news/2013-08-20/amazon-speeds-warehouse-spending-amid-ebay-delivery-threat-tech.

Langley, J. (2010) *2010 third party logistics study*. Retrieved September 20, 2013, from http://www.3plstudy.com.

Langley, J. (2013) *2013 third party logistics study*. Retrieved September 20, 2013, from http://www.3plstudy.com.

Lee, H. (2004) The triple-a supply chain. *Harvard Business Review*, 82(10), 102–112.

Lee, H., Padmanabhan, V., and Whang, S. (1997) The bullwhip effect in supply chains. *Sloan Management Review*, 38(3), 93–102.

Lieb, R., and Butner, K. (2007) The North American third-party logistics industry in 2006: The provider CEO perspective. *Transportation Journal*, 46(3), 40–52.

McCarthy, I. P. (2004) Special issue editorial: The what, why and how of mass customization. *Production Planning and Control*, 15(4), 347–351.

Mentzer, J. T. (2004) *Fundamentals of supply chain management*. Thousand Oaks, CA: Sage.

Mentzer, J. T., and Konrad, B. P. (1991) An efficiency/effectiveness approach to logistics performance analysis. *Journal of Business Logistics*, 12(1), 33–63.

Miles, S. (2013) Optimize cost and service: What supply chain visibility leaders can teach us. *Supply Chain Digest*. Retrieved September 18, 2013, from http://www.scdigest.com/experts/Amberroad_13-08-29.php?cid=7355.

Muckstadt, J. A., Murray, D. H., Rappold, J. A, and Collins, D. E. (2003) The five principles of supply chain management: An innovative approach to managing uncertainty. Cayuga Partners: Belle Meade, NJ.

Oxford Dictionary (2013a). Retrieved September 13, 2013, from http://oxforddictionaries.com/definition/english/principle.

Oxford Dictionary (2013b). Retrieved September 13, 2013, from http://oxforddictionaries.com/definition/english/strategy.

Porter, M. (1996, November/December). What is strategy? *Harvard Business Review*, 61–78.

Reuters (2013) Food fight: Wal-Mart vows to guarantee groceries, buy local. *CNBC*. Retrieved September 17, 2013, from http://www.cnbc.com/id/100784882.

Rieger, T. (2011) Overcoming barriers to success. *Gallup Business Journal*. Retrieved September 13, 2013, from http://businessjournal.gallup.com/content/145901/overcoming-barriers-success.aspx

Rosen, E. (2010) Smashing silos. *Bloomberg Businessweek*. Retrieved September 18, 2013, from http://www.businessweek.com/managing/content/feb2010/ca2010025_358633.htm.

Sheffi, Y. (2005) The resilient enterprise. Cambridge, MA: MIT Press.

Simchi-Levi, D., Peruvankal, J. P., Mulani, N., Read, B., and Ferreira, J. (2012) Is it time to rethink your manufacturing strategy? *MIT Sloan Management Review*, 53(2), 20–22.

Slone, R., Dittmann, J.P. and Mentzer, J.T. (2010), *The New Supply Chain Agenda: Five Steps that Drive Real Value*. Harvard Business Review Press: Boston MA.

Srinivasan, M. (2004) Streamlined: 14 principles for building and managing the lean supply chain. Mason, OH: Texere.

Stank, T. P., Keller, S. B., and Daugherty, P. J. (2001) Supply chain collaboration and logistical service performance. *Journal of Business Logistics*, 22(1), 29–48.

The Free Dictionary. (2013) Retrieved September 13, 2013, from http://www.thefreedictionary.com/barrier.

Thomas, K. (2012) Supply chain segmentation: 10 steps to greater profits, *Supply Chain Quarterly*, 6 (1). Retrieved September 13, 2013, from http://www.supplychainquarterly.com/topics/Strategy/201201segmentation/.

Van Hoek, R. I. (2001) The rediscovery of postponement: A literature review and directions for research. *Journal of Operations Management*, 19(2), 161–184.

Waller, M. A., Dabholkar, P. A., and Gentry, J. J. (2000) Postponement, product customization, and market-oriented supply chain management. *Journal of Business Logistics*, 21(2), 133–159.

World Commission on Environment and Development. (1987) *Our common future*. New York: Oxford University Press.

World Economic Forum. (2013) *Enabling trade: Valuing growth opportunities*. Retrieved September 16, 2013, from http://www3.weforum.org/docs/WEF_SCT_EnablingTrade_Report_2013.pdf.

4

SUPPLY CHAIN INFORMATION AND TECHNOLOGY

The supply chain principles, processes, and strategies discussed in the preceding chapters share a common bond. Each one depends on the availability of accurate and timely information. This information must be readily accessible and flow seamlessly within and across organizations to ensure that supply chain managers have the intelligence needed to make appropriate decisions. Technology is the conduit through which this information flows, making it an essential element of an integrated supply chain.

A well-planned and properly linked supply chain information system allows each organization in the channel to access a common pool of data. This fosters consistency of decision making in which activities are driven by reality rather than forecasts and opinions. For example, an integrated supply chain information system can capture and distribute point-of-sale (POS) data to the retailer, manufacturer, and components suppliers. These daily demand signals facilitate effective decision making about store replenishment, distribution center reorders, production schedules, and materials requirements. Also the bullwhip effect is minimized because this information-driven supply chain is not affected by independent forecasts and the associated forecast errors.

The value of capturing real-time information and making it visible to decision makers across the supply chain is well recognized. Organizations are actively investing in supply chain information systems to enhance the customer experience and support growth. A recent study of supply chain technology users found that 60 percent of the respondents plan to increase their investments in supply chain technologies over the next few years (Klappich, 2012). These technology purchasers prioritize holistic order-to-delivery solutions that are supported by enterprise applications. Collectively, they provide greater transparency within the supply chain and help the organizations optimize their logistics and distribution operations (PwC, 2013).

The constant evolution of supply chains underlies the need for rapid information dissemination and technology investments. As supply chains become more complex and demand driven with tailored products, omnichannel fulfillment options, and faster speed to market, managers must adapt their supply chain information systems. To this end, there is a move toward more sophisticated tools to manage complex supply chain activities, support visibility requirements, and synchronize supply chain flows (Holcomb, Nightengale, Ross, and Manrodt, 2011).

Maximizing the contribution of supply chain information systems to organizational success depends upon a clear technology strategy and investment in tools that will support current and future needs. This chapter begins with an examination of supply chain information requirements, value, and challenges. Next, technology considerations are discussed along with a discussion of supply chain information system elements. An in-depth review of key supply chain software tools and supporting technologies follows next. The chapter closes with discussion of future issues, emerging capabilities, and innovations that will drive supply chain information systems development.

By the end of this chapter, it will become evident that effective supply chain information systems and software are vital for creating responsive strategies, synchronizing plans, and executing processes to efficiently and effectively serve customer requirements.

Information Needs in the Supply Chain

Information drives nearly every decision and process in the supply chain. Decisions about product replenishment are based on POS data, transportation mode selection is driven by customer delivery requirements, purchase decisions are derived from inventory level reports, and so on. In essence, information is the glue that holds the supply chain together and facilitates the planning and execution of an integrated supply chain. Without this information bond, managers lose their virtual line of sight to inventory, demand, and activities taking place at supplier and customer locations. Blind spots result, and opportunities for collaboration are lost, leaving decisions to be based on educated guesses and internal signals rather than chain-wide knowledge.

Information Needs

Supply chains are information-intensive, and managers at every level require information for their key activities. According to Bozarth and Handfield (2013), this includes information for strategic decision making, tactical planning, routing decision making, and execution and transaction processing.

Strategic decision making focuses on the development of long-range supply chain plans that align with the organization's vision and mission. The information needed is often unstructured and may differ from one project to the next. For example, a wide variety of supply, demand, and operational cost data must be captured from a variety of sources for a strategic network design project. In contrast, a new product development decision will require design, capacity, and supply capability information. This type of information is used to evaluate strategic alternatives and conduct what-if analyses using decision support tools.

Tactical planning focuses on the linkages between organization and coordination of activities in the supply chain. The information must be readily available, support planning processes, and be in a flexible format that can be modified by the various supply chain participants for use in their systems. For example, sales and operations planning (S&OP) requires sharing of information about demand patterns, promotional plans, supply capacity, inventory, and related data to create a unified operating plan. This helps each group optimize its plans through the allocation of critical resources including people, materials, capacity, money, and time resources (Hitachi Consulting, 2007).

Routine decision making requires operational-level information that can be captured for rules-based decision making. The information needs to be fairly standardized so that it can be easily used by a system to generate standardized solutions. For example, an automated transportation routing guide requires information on a shipment's origin, destination, product characteristics, weight and dimensions, and service level requirements to recommend an appropriate mode and carrier. Decision makers retain the ability to review the recommended solution before action is taken and deviate if needed.

Execution and transaction processing uses basic information from supply chain databases, customer profiles, inventory records, and related sources to complete fulfillment activities. The information must be accurate, available for retrieval in a timely fashion, and standardized so that it can be processed in a highly automated fashion. For example, information from an online customer's order should be able to be captured, compared to available inventory levels, and scheduled for processing without human intervention. This will support efficient and rapid fulfillment of the order.

Information Flows

The broad scope of supply chain planning and decision making affects not only information needs but also who needs the information and how it flows across the supply chain. Internal information flows are needed to support cross-functional collaboration and optimize organization-wide performance. For example, marketing, operations, finance, and logistics professionals must all contribute information to and derive information from the S&OP process for it to generate key benefits. A failure to establish internal information flows will lead to functional silos, myopic decision making, and suboptimal performance.

Information must also flow seamlessly between the organization and its supply chain partners to promote integrated decision making and process synchronization. A consistent stream of customer demand information is needed to drive effective upstream production and purchasing decisions by manufacturers and suppliers. Shared insights regarding supplier capacity, production schedules, and inventory availability will facilitate alignment and efficient execution of downstream processes.

Logistics service providers must also be kept in the upstream and downstream information loops regarding demand, inventory levels, and delivery schedules. This knowledge allows the service providers to marshal the people and equipment resources needed for timely product flows. In addition, financial institutions participate in the movement of relevant information regarding payments and transactions, while government agencies require ongoing communication regarding trade data and regulatory compliance. Figure 4-1 highlights the essential flows of supply chain information between the supply chain participants.

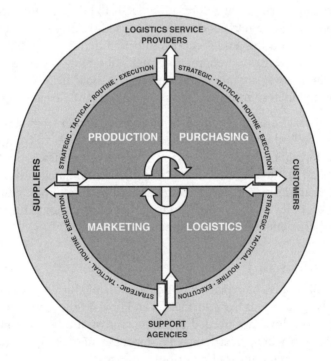

Figure 4-1   Supply chain information flows

Information Characteristics

Information flows between partners will be of value if the transfer provides quality data and actionable knowledge. The garbage-in-garbage-out axiom of information technology holds true because supply chain information systems cannot overcome poor quality information. Inferior information quality often leads to improper decisions and unanticipated results. Hence, the information needs to be of high quality to support effective analysis and decision making.

There are many important characteristics of information quality, including accuracy, accessibility, relevance, timeliness, and transferability (Coyle, Langley, Novack, & Gibson, 2013). Riley (2012) adds the important considerations of reliability and value, whereas Chopra and Mendl (2013) identify the need for usability and a shared view. Each is discussed in the following sections.

Accuracy

Supply chain information must be correct and depict reality. A true picture of the supply chain that is free of errors begins with proper data collection and facilitates logical decision making. Otherwise, information inaccuracies will hamper a manager's assessment of the situation and may lead to inventory shortages, transportation delays, dissatisfied customers, and financial penalties.

Accessibility

Accurate information must be readily available to supply chain managers to perform their duties. This sounds simple enough with the prevalence of Enterprise Resource Planning (ERP) systems and the emergence of cloud computing, but supply chains present unique challenges. Supply chain information may be controlled by external organizations, distributed among multiple locations, and contained in disparate systems. Technical challenges and trust issues must be overcome to provide accessibility to individuals who have a legitimate need for information.

Relevance

Supply chain managers must have access to pertinent information for analysis and decision-making purposes. The information must be void of unnecessary and excessive detail because the irrelevant data will slow progress, shroud important issues, and overwhelm decision makers. They would be better served with succinct and tailored information that comes from customized information queries, exception reports, and segmentation analysis.

Timeliness

Information is relevant only if it is up-to-date and available in a reasonable timeframe. Highly synchronized and demand-driven supply chains depend upon the availability of real-time data that provides cross-channel visibility and responsiveness. With timely information flows, activities can be monitored, and managers can quickly respond to exceptions with corrective actions. This will prevent minor problems from escalating into full blown crises.

Transferability

To facilitate timely and relevant access to information and collaborative decision making, information must be readily transferable from one location to another. Fortunately, the Internet and cloud computing platforms make information transfers relatively easy, inexpensive, and safe, though organizations must take precautions to ensure the security of sensitive data.

Reliability

The information presented in reports and transaction data sets must come from reliable and authoritative sources. The data provided must also be reasonably complete, accurate, and unaltered to meet the intended purposes. When incomplete or estimated data is provided, a clear explanation of the missing values and assumptions is needed so that a supply chain manager can adjust his analysis accordingly.

Value

Generating reliable information is not a cost-free proposition. The hardware and software technologies needed to capture and manage supply chain data can be expensive. In fact, the average supply chain software spent in 2012 for licensing, integration, and training was $553,400 among respondents of a Peerless Research Group survey (McCrea, 2012). Buyers must determine whether these investments truly enhance knowledge and produce tangible performance benefits.

Usability

Information is valuable only if it can be used to make better decisions. Efforts must be made to define information requirements and capture appropriate data. Time must not be wasted capturing extraneous data that is not usable. Also, information is usable only if it can be seamlessly shared and translated from one format to another with no discernible loss of data.

Shared View

Supply chain managers and their counterparts must share a common perspective regarding what information will be used to make decisions and how those decisions will be made. If different functions and stakeholders operate independently using standalone information, then synchronization opportunities will be lost and poor performance will occur.

Collectively, these nine characteristics generate quality information for efficient and effective decision making by supply chain managers. That is why organizations invest in supply chain information systems and spend significant effort linking them to other internal functions, suppliers, and customers.

Technology Considerations

The potential value of information technology is well recognized by supply chain managers. They understand that technology plays a critical role in efficient execution of the supply chain processes described in Chapter 2. Managers also realize that an effective supply chain information system will help them embrace the guiding principles and pursue the innovative strategies discussed in Chapter 3.

These beliefs are well founded. Multiple studies confirm that supply chain information technology contributes to supply chain success. Premus and Sanders (2005) found that technology has a direct and positive impact on organizational performance as well as internal and external collaboration. Accenture (2009) found that high-performing organizations deploy supply chain information systems that produce insightful analytics, alignment, and responsiveness. And Holcomb et al. (2011) indicated that technology enables organizations to boost supply chain flexibility, efficiency, and differentiation capability. Ultimately, a well-planned and executed technology system helps an organization and its stakeholders optimize global performance and mitigate risk (Simchi-Levi, Kaminsky, & Simchi-Levi, 2008).

Information Technology Capabilities

Technology capabilities are vital to an organization's business survival because an effective and efficient system with 24/7 availability is essential for competing in the marketplace (Ernst & Young, 2011). Not only must these capabilities support competitive plan, buy, make, move, and reverse processes but they must also enable organizations to pursue strategic supply chain initiatives. As new strategies emerge, the technology capabilities of the supply chain information system must evolve.

There are numerous technological capabilities needed to facilitate supply chain process efficiency and strategy effectiveness. Information systems must effectively support optimization, cross-chain visibility, speed to market, agility, collaboration, adaptability, differentiation, and risk management. Without properly configured technological support, each of these goals is little more than an unrealized ambition.

Optimization

Peak performance of a supply chain requires an ideal alignment of resources within and across organizations. There are many options and trade-offs to consider when analyzing opportunities to reduce costs and improve service. Supply chain optimization technologies use mathematical modeling tools to quickly run through the options to find the solution that facilitates success for all supply chain stakeholders. Organizations can use these tools to study network design options, determine appropriate inventory levels, develop routing decisions, and more. The goal is to maximize service for the minimal possible operating costs.

Cross-Chain Visibility

Managers need up-to-date awareness of, and control over, key supply chain activities. Having the most current data about the supply chain is a prerequisite for effective decision making by managers regarding supply chain events. Visibility tools capture valuable real-time data about the current state of the supply chain, filter it, and put it into an accessible format for stakeholders. They can use the knowledge to close the loop between planning and execution, synchronize end-to-end activities, and quickly respond to exceptions (Heaney, 2013).

Speed to Market

The velocity at which the end-to-end flow of product occurs in the supply chain must conform to customer requirements. It will vary based on the situation—normal versus emergency replenishment situations, new product introduction versus existing product flows, and so on—and organizations need the capability to adjust speeds accordingly. Properly implemented technologies capture these requirements, prioritize them, and determine delivery method to ensure that the speed to market is consistently aligned with customer needs, be it next-day, next-week, or next-month fulfillment.

Agility

In complex, uncertain, and fast-changing markets, supply chain managers must have the ability to recalibrate plans and quickly respond to volatility in supply and demand. Agile and coordinated supply chains have the capability, capacity, and flexibility to deliver the same or comparable cost, quality, and customer service under changing conditions. Appropriate technology has strong decision support analytics with the ability to model scenarios to understand volatility impacts and help the organization sense, translate, and respond to changes in demand and supply (Cecere, 2012).

Collaboration

As stated previously, information accessibility and a shared view of information are necessary to promote consistency and coordination in the supply chain. Managers can effectively collaborate on strategies, processes, and exception response only when they have access to the same data in a timely fashion. Technology provides the electronic linkages across geographically dispersed supply chains that facilitate seamless information sharing. The Internet, extranets, virtual private networks, and cloud computing support collaboration capabilities.

Adaptability

Managers must adapt the design and capabilities of a supply chain to evolving market conditions. Organizations can capitalize on demographic trends, political shifts, emerging economies, and other new opportunities through alteration of their supply chain operating model. This requires a flexible, geographically dispersed network supported by strong technology to analyze options and determine profitable allocation of network capacity. By linking supply chain technologies to sales and marketing systems, companies can sense and respond to real-time market needs and shape demand when capacity is limited (Cudahy, George, Godfrey, and Rollman, 2012).

Differentiation

Organizations must dynamic align their demand and supply response capabilities to optimize net profitability across each segment. By segmenting customers and offering differentiated service levels, the organization can increase sales and reduce costs. They avoid the "one size fits all" strategy that underserves important customers and creates unnecessary costs (Holcomb et al., 2011). Technology can help an organization define supplier and customer segments, understand the cost to serve them, and prioritize service execution so that the critical groups receive appropriate attention.

Risk Management

Efforts to generate supply chain efficiencies through lower inventories, offshore manufacturing, and rationalized supplier bases can make the supply chain inflexible and more prone to risk. This, combined with economic volatility, shorter product life cycles, and other challenges makes it critical for managers to address supply chain risk. Being able to distinguish among and between risks is a key capability for success. Organizations should use predictive and real-time risk management technologies, in combination with scenario planning and execution technologies, to drive far greater levels of supply chain dynamism (Cudahy et al., 2012).

This list of capabilities is by no means comprehensive. Information technology must also support initiatives related to supply chain synchronization, innovation, performance analysis and improvement, and profitability. Hence, the capabilities of an information system must be regularly revisited and upgraded to support dynamic supply chain requirements.

Information Technology Challenges

Technology holds great promise for enhancing supply chain performance and organizational competitiveness. However, technology only enables success rather than guaranteeing it. Supply chain managers must evaluate their technology options with caution and understand the potential challenges and pitfalls that exist when adopting new tools or updating new ones.

A fundamental challenge is the unrealistic assumption that supply chain technologies will fix everything. Managers must remember that technology is a facilitating tool rather than a silver bullet solution. Technology cannot overcome poorly trained talent, improperly designed processes, or inaccurate data. These challenges must be resolved prior to technology adoption.

Also, for technology to work most effectively, a company must have an explicitly stated and shared corporate vision and mission—not just among external customers and service providers but also internal constituents involved in the supply chain, including warehousing, transportation, and sales. Furthermore, supply chain management technologies are most effective when they support tight collaboration among these parties and provide visibility into all key aspects of the business (Brady, 2013).

Another challenge is the inconsistent use of supply chain technology within and across organizations. In an Aberdeen Group study, large companies reported that their international supply chains are only 50 percent as automated as their domestic supply chains. Overall, only 6 percent of companies reported that they have highly automated end-to-end and cross-functional processes (Enslow, 2006). To fully benefit from the power of technology, organizations should create stable enterprise-wide platforms using a uniform set of supply chain software. This will ensure that data can readily flow between managers for accurate analysis and informed decision making.

Despite significant efforts and improvements, systems integration continues to be a stumbling block for many organizations. Chief information officers at logistics service providers view integration with customers' information technologies as their single biggest challenge (Reynolds & Khan, 2013). There are also integration challenges on the customer side that limit the impact of technology. On the customer side, supply chain network complexity and difficulties in creating visibility across the supply chain are top challenges according to a recent Gartner study (Pearson Specter, 2013).

Poor implementation practice is another technology pitfall that continues to cause problems. A failure to create a change management plan with a staged, logical approach to adopting new technologies increases the risk of implementation delays, lost connectivity, and supply chain disruptions. Other organizations fail to prepare employees for the new technology. Limited training may lead to suboptimal use of technology as employees do not understand the full array of software features and capabilities. Finally, some organizations do not establish adequate budgets for technology installation and implementation (Coyle et al., 2013).

Left unaddressed, these challenges will limit opportunities to achieve the desired return on investment (ROI) for supply chain technology purchases. In a recent study by Peerless Research Group, survey respondents pointed to several of these obstacles (Quinn, 2013). Figure 4-2 reveals that cultural change related to implementation led the list of ROI obstacles. Lavish spending on technology will yield limited results unless these issues are properly addressed.

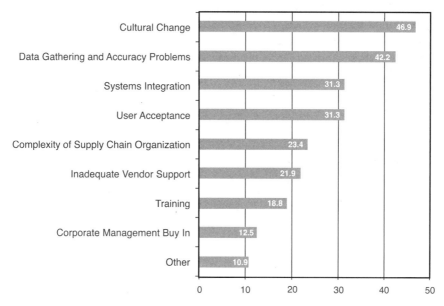

Figure 4-2   Return on investment obstacles (Source: Quinn, F. 2013, "Maximizing Your Return on Investment from Supply Chain Technology" and Supply Chain, 247.)

It is important to note that these challenges are not insurmountable. Many organizations have successfully adopted and implemented supply chain technologies that facilitate cost control, visibility, and service improvement. The key is to view technology implementation as a business improvement project rather than an information technology project. Supply chain leaders must take an active role in the planning, implementation, and evaluation of the new tools rather than assign full responsibility and control to information technology staffers, software vendors, or outside consultants.

Information Technology Framework

Overcoming technology challenges can be accomplished through its purposeful inclusion in an overall supply chain management framework. When technology is linked to people and processes in an intentional and integrated fashion, many of the oversight issues that can plague an implementation are avoided. People will get the necessary training to build important software skills and processes will be updated to gain full advantage of the new technology tools. The outcome will be a smoother introduction of the technology and greater benefits for the organization.

The Master Model of Supply Chain Excellence (Moore, Manrodt, & Holcomb, 2005) provides a sound basis for an information technology framework. In addition to the essential elements of skilled people, strong processes, and integrated technology, the framework contains key technology requirements and differentiating capabilities. Figure 4-3 depicts the integration of these elements into a logical framework that can be readily adopted by supply chain managers.

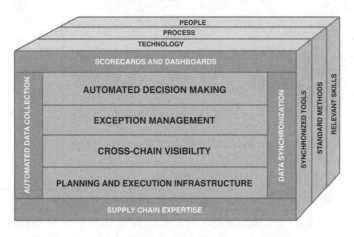

Figure 4-3   Information technology framework

By themselves, software and other supply chain technology components cannot provide actionable knowledge for supply chain managers. Data must be collected using automated capture tools when possible and synchronized across the supply chain. Standardized, complete, accurate, and consistently aligned data allows managers to successfully plan and execute supply chain processes. Computer-based scorecards and dashboards allow organizations to continuously monitor and analyze performance. The technology can alert managers to performance exceptions and sometimes recommend process adjustments. Finally, functional expertise is needed to glean knowledge from the supply chain technology to support planning and decision making across all supply chain operations—procurement, production, delivery, and returns.

After the foundation has been built and the key requirements attained, supply chain stakeholders must integrate processes and improve connectivity of their information systems to establish differentiated capabilities. A strong planning and execution infrastructure, consisting of supply chain software tools, provides cross-chain speed, optimization, and connectivity. Visibility tools provide a seamless flow of timely, important information across the supply chain that enhances intelligence and shifts supply chain managers from reactive to proactive strategic planning, collaboration, and decision making. Exception management capabilities leverage visibility to rapidly detect performance problems, alert the affected organizations, and trigger corrective actions. Automated decision making is the pinnacle of differentiating capabilities, using the exception management data and preestablished rules to evaluate events, formulate potential responses, and recommend solutions.

The foundation elements, key requirements, and differentiating capabilities of the information technology framework must be developed in a logical, sequential fashion. Patience is critical as it will take time to integrate systems, synchronize data, and institutionalize cross-chain collaboration before significant benefits are realized. For those who commit time and resources to a cohesive technology framework that translates the vast amounts of supply chain data being generated today into actionable intelligence, the payoffs will be significant.

Supply Chain Software

An essential component of the supply chain framework is the software tools that allow managers to arrange, analyze, analyze, and act upon relevant data. The software market space includes supply chain technologies that address virtually every strategy and operation related to the synchronization of supply and demand. Whether the organization needs to design a supply chain network, develop a demand forecast, optimize global transportation, maintain inventory visibility, or monitor multichannel fulfillment activities, there is relevant software available These tools harness the computational power and communication abilities of today's technology to help organizations plan, execute, and control fulfillment activities in real time.

Organizations readily recognize the potential benefits facilitated by supply chain software and have been spending heavily to expand their information capabilities. The worldwide supply chain management software market totaled $7.7 billion in 2011, a 12.3 percent increase from 2010, according to Gartner, Inc. It was the second year of double-digit growth for the SCM software market as organizations continued to prioritize spending on supply chain applications. About one-half of this spend was made with five solutions providers, with smaller players capturing a significant chunk of the business in this highly fragmented software market (Gartner, 2012). Table 4-1 identifies the primary supply chain solutions providers.

Table 4-1 Top 10 SCM Solutions Providers

Rank	Company	2012 Revenue	Planning Software	Execution Software
1	SAP	$1.7 billion	✓	✓
2	Oracle	$1.5 billion	✓	✓
3	JDA Software	$426 million	✓	✓
4	Manhattan Associates	$160 million	✓	✓
5	Epicor	$138.2 million	✓	✓
6	IBM	$112 million	✓	
7	Infor Global Solutions	$111 million	✓	✓
8	RedPrairie*	$105 million		✓
9	Descartes Systems Group	$96 million		✓
10	Kewill Systems	$62 million		✓
10	Unit4	$62 million		✓

* Merged with JDA Software in late 2012

Adapted from: Trebilcock, B. (2013.) Top 20 SCM software suppliers, 2013. *Modern Materials Handling*, 68(7), 32–36.

This spending goes toward a broad range of software solutions to achieve the supply chain goals discussed in Chapter 1. Figure 4-4 presents a logical categorization of supply chain software tools and the related systems that create a stable foundation for an integrated system. Each category of software is discussed, with a focus on its general role in the supply chain and its intended contribution to performance.

It should be noted that maximum impact cannot be achieved by investment in a single software category. A coordinated approach to software selection and implementation from these categories is necessary to create a well-linked supply chain information system. Doing so will help an organization drive supply chain efficiency, customer value, responsiveness, resiliency, and financial success.

Figure 4-4   Supply chain software

Supply Chain Planning

Supply chain planning applications encompass a comprehensive set of software tools designed to help managers gain more accurate, detailed insight into issues that affect their planning and development of supply chain activities. The solutions use complex algorithms, optimization techniques, and heuristics to solve supply chain objectives within the stated planning horizon (ARC Advisory Group, 2012).

The overall goal of a supply chain planning software implementation is to move from manual, autonomous planning processes to software that helps organizations leverage real-time data and promote collaboration across departments, suppliers, and customers for synchronized planning. Ultimately, stronger planning capabilities will facilitate better operational and tactical decisions, leading to more efficient process execution, reduced waste and stockouts, and improved profitability. A strong planning system will allocate resources, develop forecasts, and establish schedules to support demand fulfillment.

The capabilities of these software tools are becoming more critical to business success. One reason is because supply chain planning is getting more complex. Organizations are taking on new planning challenges that require more powerful computing capabilities, such as understanding omnichannel demand, forecasting by stock-keeping unit (SKU) at the store level, and managing multiple plans simultaneously across segmented supply chains. Also, integration of planning and execution must improve. Creating feedback loops of execution data back into planning systems will provide more robust information for planning purposes (SCDigest Editorial Staff, 2013).

According to the Gartner IT Glossary, a supply chain planning suite sits on top of a transactional system to provide planning, what-if scenario analysis capabilities, and real-time demand commitments within the context of existing constraints. Typical supply chain planning modules include the following:

- Strategic network design
- Master planning

- Sales and operations planning
- Procurement planning
- Production planning and scheduling
- Demand planning
- Distribution planning
- Inventory planning and optimization
- Vendor-managed inventory
- Capacity planning

Used in an integrated fashion, as highlighted by Figure 4-5, these solutions enable a supply chain manager to view, analyze, simulate, and segment essential data. This provides a clear vision of alternatives so the decision maker can make the most effective plans. For example, one software provider's supply chain planning suite consists of 20 different applications that collectively focus on providing global, end-to-end supply chain optimization across procurement, manufacturing, and distribution. This enables users to increase productivity, improve service levels, reduce operation costs, and drive profitable growth. Prominent planning tools are discussed in the following sections.

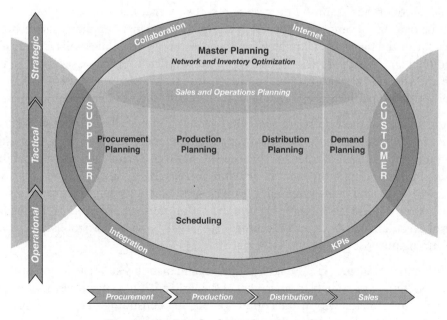

Figure 4-5   Supply chain planning software

Source: *n.SDKEP Supply Chain Solutions Suite*. Retrieved from http://www.dys.com/en/solutions/supply-chain-solutions

Master Planning

Master planning focuses on the synchronization of purchasing, production, and distribution activities for timely fulfillment of customer demand. The master plan efficiently allocates supply chain resources based on established business priorities to minimize system-wide cost or maximize profit over multiple time periods.

This software overcomes the time-consuming processes and delays of a manual planning system that cannot respond to disruptions or demand shifts in a timely fashion. Instead, a master planning solution creates actionable purchasing, production, and replenishment orders, provides increased visibility for potential problems in the supply chain, and suggests potential resolutions. In addition, inventory levels needed for smooth material flows are reduced, and capital-intensive resources are utilized more efficiently, yielding more output from the supply chain.

Network Design

Network design software provides a fact-based analysis of existing and proposed supply chain facilities to help organizations determine how to most efficiently serve customers, consolidate new assets, or expand into new markets or products. These solutions evaluate alternative network options to determine the lowest cost supply chain configuration that satisfies demand at a particular service level.

Using these network design solutions, users can model and simulate complex supply chains. At the strategic level, the software helps organizations determine the right number, location, and necessary capacity of production and warehousing facilities. Tactical level capabilities ascertain product sourcing and production locations, as well as assign customers to facilities. Finally, operational level analysis focuses on how product should flow through the network (Banker, 2012).

Sales and Operations Planning

Sales and operations planning (S&OP) is an iterative management process that determines a single operating plan to profitably synchronize supply and demand in the context of the organization's strategic objectives. The process is built upon stakeholder agreement and an approved consensus plan. An effective S&OP software application will help the organization optimize production costs and plan its demand down to the part level from multiple tiers of suppliers. These applications also incorporate financial measures to promote top-line revenue growth and overall profitability.

An S&OP system contains preconfigured processes and key performance indicators (KPIs) to create a framework for balancing demand and supply. To ensure an accurate action plan based on real-time data, these tools derive key information from current demand planning, sales forecasting, and production planning initiatives. The S&OP software also uses dashboards that display data related to equipment, labor, facilities, material, and finance, providing stakeholders with a single, shared view of the data. Finally, S&OP software allows managers to analyze what-if scenarios to determine the probable impact of new strategies, supply chain disruptions, and other events.

Procurement Planning

Procurement planning is the process of deciding what to buy, when it will be bought, and from which supplier. Procurement planning software facilitates the process by helping the organization identify all requirements expected to be sourced over a period of time and developing procurement schedules. The goal is to ensure component and raw materials supply availability from optimal suppliers for implementing the production plan.

Procurement planning software enables the organization to unify and control strategic sourcing while maintaining site-level autonomy for operational decision making flexibility. These tools provide a variety of decision support tools for procurement, including analyses of alternative sourcing deals, time-based replenishment recommendations, optimal safety stock and service levels, and order batching to achieve transportation discounts. Effective solutions reduce the time and cost to successfully administer complex, fast-moving procurement processes.

Production Planning

Production planning strives to ensure that sufficient raw materials, components, and other necessary items are available to manufacture products according to the schedule specified that satisfies customer demand over a specified time horizon. The purpose of production planning software is to minimize production time and cost via more efficient use of resources. It seeks to optimize the master production schedule according to demand, inventory levels, plant capacity, and related constraints.

Production planning software helps the organization monitor, manage, analyze, and report on the end-to-end manufacturing process. These tools help organizations model production plans, automate the planning process, and achieve real-time visibility and control. Using simultaneous detailed capacity planning and materials planning, key modules rapidly assess an organization's capability to promise status, create materials requirements plans, and define the master production schedule. An advanced planning and scheduling module supports real-time production schedule adjustments to allow for unplanned events while maintaining optimal throughput. Ultimately, a capable production planning solution prevents unnecessary downtime while increasing manufacturing efficiency, quality, and profit margins.

Demand Planning

Demand planning is a multistep process intended to generate reliable demand forecasts. Key activities in the process include developing front-end agreements and planning horizons by stakeholders, creating sales forecasts based on historical data, collaborating on exceptions, building order forecasts, and establishing the consensus demand plan. Effective demand planning software helps users improve the accuracy of sales and order forecasts, align inventory levels with peaks and troughs in demand, and enhance profitability for a given channel or product.

Demand planning applications attempt to provide an accurate, global view of reality to help managers predict and shape demand. These tools can integrate seasonality, promotional events, and product life-cycle changes into the analysis of demand and subsequent forecasts. The use of best-in-class demand planning systems results in higher forecast accuracies and lower inventories, according to an Aberdeen Group study (Blanchard, 2008).

Distribution Planning

Distribution planning encompasses a wide variety of activities that focus on ensuring inventory availability across the supply chain. Through these processes, managers make use of relevant information to determine when and in what quantities certain items must be replenished in order to maintain adequate inventory levels for efficient production and order fulfillment. The goal is to achieve demand driven replenishment that avoids out-of-stocks, minimizes spending for safety stock and expedited transportation, and reduces risk of product obsolescence.

Distribution planning software provides a tactical tool for optimizing the day-to-day flow of inventory between stakeholders across the supply chain. Distribution requirements planning tools establishes orders based on actual daily consumption, remaining inventory levels, production capabilities, and supply chain constraints to synchronize inventory levels across the supply chain. Product availability is increased while overall inventory, ordering, holding, and transportation costs are minimized. Other distribution planning software focuses specifically on the delivery component to optimize delivery routes, fleet efficiency, and transportation costs.

Supply Chain Execution

The recommendations and decisions generated by supply chain planning systems are carried out by supply chain execution applications. The software facilitates the performance of day-to-day operating tasks from the time an order is placed until the delivery is made. These tools direct manufacturing operations, control fulfillment processes, and guide transportation flows in an attempt to make the supply chain operate smoothly and cost efficiently.

Interest and investment in execution tools is growing because of their rapid ROI and positive effect on the supply chain. Successful implementation of supply chain execution applications provides users with better control of costs, stronger customer service, greater inventory visibility, improved data accuracy, faster throughput, and higher inventory turns (Maloney, 2006). A recent survey of chief information officers at logistics service providers indicates that these execution software benefits continue to be a primary reason why organizations are investing in supply chain applications (Reynolds & Khan, 2013).

In 2012, the estimated global spend for supply chain execution software was nearly $2.5 billion. Future outlays are expected to exceed $4.1 billion by 2017. Nearly one-half of the current spend is on warehouse management systems, followed by transportation management systems (Trebilcock, 2013). Still, there is much room for growth as adoption rates are less than 50 percent for most execution tools, according to a Peerless Research Group study (McCrea, 2013d). Future spending activity will continue to concentrate on these two execution applications, as revealed in Figure 4-6.

Figure 4-6 Software purchase and upgrade plans in the next 12 months

Source: McCrea, B. (2013d.). 11th annual software users survey: Caution remains. *Logistics Management* 52(6): 36–40.

Traditionally, execution tools were standalone applications that focused on internal logistics activities such as order management, warehouse management, inventory management, and transportation management. With attention shifting to integrated fulfillment capabilities, the category now encompasses a broader array of functionality, including global trade management, manufacturing execution, and distributed order management. By implementing integrated suites, organizations benefit from these supply chain execution applications working together and using common data to synchronize and optimize the processes necessary to serve customer demand (Rogers, 2011). Capabilities and uses of the primary supply chain execution tools are discussed next.

Warehouse Management Systems

Warehouse management systems (WMS) are widely used to coordinate the flow of materials and information throughout the fulfillment process. WMS is a software control system that improves warehouse operations through efficient management of information and completion of warehouse tasks, with a high level of control and inventory accuracy. These tasks include receiving, put-away, picking, packing, shipping, storage location, work planning, warehouse layout, and analysis activities (Coyle et al., 2013).

WMSs are widely used because they provide rapid improvement of fulfillment operations in terms of labor efficiency and productivity, inventory accuracy, and facility utilization. Labor performance improves as the WMS optimizes work paths, interleaves activities, and provides explicit inventory location instructions. By maintaining an accurate database of inventory levels, location, and expiration dates, inventory can be readily controlled. Also, the use of verification technology helps workers fulfill orders correctly, which avoids on-hand overages and shortages. The WMS also improves space utilization by determining the optimal storage patterns to maximize space utilization. Finally, a WMS will help the organization reduce lead times and free up working capital by accelerating the flow of merchandise.

In addition to the execution benefits, WMS applications provide tactical planning and reporting capabilities to improve managerial control of fulfillment operations. Product velocity and slotting data can also be used to reconfigure and optimize the layout of products in a facility as business requirements change. Integration with labor management tools allows the WMS to plan and balance labor requirements. The WMS also allows managers to create cross-docking and flow-through capabilities for demand spikes and promotional orders. WMS performance measurement capabilities support the analysis of fulfillment operations and process improvement planning.

Though many organizations use a WMS as a point solution for managing individual fulfillment sites, innovators link it to other planning and execution tools to enhance supply chain visibility, accelerate flows, and control costs. Production schedules determine storage space requirements, inbound delivery information drives receiving schedules, and distributed order management tools review WMS databases to determine order fulfillment location based on costs and inventory availability. Finally, advancements in wireless technology and automated identification are fueling the growth of paperless fulfillment, integration of materials handling equipment, and other performance-enhancing initiatives (Napolitano, 2012).

Transportation Management Systems

The planning and flow of materials across the supply chain are handled by transportation management systems (TMS). A simple definition of a TMS focuses on the booking, execution, and settlement of transportation movements with a focus on optimizing cost and speed. However, a TMS is more than freight management software. Current TMSs provide hybrid planning-execution-evaluation capabilities to assist managers in nearly every aspect of transportation, from basic load configuration to complex transportation network optimization.

The planning capabilities of the TMS assist supply chain managers with pre-shipment decisions related to network design, fleet planning, rate analysis and contracting, mode and carrier selection, and routing and scheduling. The optimization capabilities of the TMS allow managers to rapidly assess all possible delivery options versus the hours and days that it would take to manually develop a transportation plan that considers only a finite number of options. Up-to-date information from the organization's WMS, demand forecasts, and customer orders can be used to make more effective transportation decisions.

TMS execution tools help transportation managers improve decisions related to shipment activities such as load tendering, in-transit tracking, delivery appointment tracking, and freight audit and payment. The TMS should enable managers to seamlessly interface with order management and WMS applications to quickly identify transportation needs based on order size, origin, destination, and service requirements. By automating many of these execution activities, a TMS helps an organization reduce errors and delays, improve accountability and visibility, and limit deviation from standard operation procedures. The result will be lower-cost, higher-quality delivery processes.

TMS evaluation tools give organizations the post-shipment capability to rapidly analyze delivery metrics related to timeliness, cost, safety, and customer satisfaction. Much of the needed information is captured directly by the TMS; other data comes from other systems and delivery documents. A strong TMS has the capability to coalesce and present the data in performance reports and scorecards that can be used for carrier benchmarking, corrective action plans, and future purchase decisions.

With more organizations attempting to squeeze efficiencies out of their supply chains by optimizing transportation, the TMS market is enjoying double-digit annual growth. Gartner estimates that TMS spending grew by 14 percent in 2012 over the previous year (Trebilcock, 2013). Part of the growth is attributable to the evolution of TMS capabilities to include private fleet options, software vendor integration of carrier information into their TMS, and enablement of multilegged, multimodal global deliveries. Also, the availability of cloud-based TMS options that improve affordability and reduce implementation time is attracting small- and mid-sized shippers to the technology (McCrea, 2013b).

Manufacturing Execution Systems

A manufacturing execution system (MES) is a control system for managing work-in-process on the factory floor. Much like a WMS, this execution tool tracks the progress and completion of production orders, manages the workload, and collects data for reporting purposes. The MES also electronically dispatches orders to personnel, adjusts the schedule as need to demand changes, and restores production flows after equipment breakdowns or quality problems occur. (Rogers, 2011)

The MES provides a number of valuable benefits that justify the investment. These applications facilitate shorter cycle times through higher facility throughput, reduce costs through enhanced product quality, and avoid nonconformance penalties through adherence to strong compliance and governance rules. The MES help managers measure every process to find new ways to bring in process improvements that can save time, money, and raw materials during manufacturing, as well as conduct detailed analyses to find increased productivity. The MES can also be tied to other reporting systems, including ERP systems and business process management applications to give companies a deep overall view of its production capabilities (Weiss, 2011).

These MES benefits are driving adoption. Gartner estimates that the market for MES applications is $1.5 billion and growing. Also contributing to the adoption are better integration of the MES with other supply chain planning and the execution tools to provide end-to-end supply chain visibility and out-of-the-box functionality. Much less custom coding is now required to install the MES, which means faster implementation at a lower investment (Trebilcock, 2013).

Global Trade Management

Managing order fulfillment and delivery of international orders is a complex undertaking, as you will learn in Chapter 6. It involves government compliance, tariffs and related taxes, INCOTERMS, and managing extended supply chains with visibility. To surmount these challenges, organizations turn to global trade management (GTM) for help. These solutions avoid paper-based, manual intervention processes, help organizations reap the benefits of free trade agreements, and achieve resiliency (Bordner, 2013).

These highly automated software systems connect cross-border trading partners electronically to provide stronger import and export processes, global logistics execution, and trade agreement control. GTM benefits include more effective collaboration and messaging, lower in-transit inventory levels, lower expediting expenses, reduced broker fees, reduced cargo insurance costs, and a variety of financing benefits. Exports gain benefits equivalent to 1.7 percent to 2.4 percent of net sales while importer gain benefits equal to 0.6 percent to 2.2 percent of net sales (Hausman, Lee, Napier, & Thompson, 2009).

Distributed Order Management

Distributed Order Management (DOM) software is a central order orchestration hub for companies with complex fulfillment networks (Banker, 2013). Retailers, in particular, gain the ability to capture, manage, and optimize orders regardless of origin—computer, retail store, kiosk, or mobile phone—via DOM applications. After order capture, the DOM software goes to work, determining which fulfillment facility or channel will best serve the customer. This decision is based on inventory availability, delivery cost, order processing capacity, customer requirements, and other factors. When implemented and executed effectively, DOM helps retailers meet service requirements through the most efficient means possible.

A well-deployed DOM system provides numerous customer and retailer benefits. The customer is not constrained by local product offerings or inventory levels, can monitor order and shipment status, and has the flexibility and convenience of returning goods to any location in the network, regardless of the order fulfillment point. The retailer can better serve demand without additional inventory or facility investment and, like the customer, gains cross-channel inventory visibility and access. Finally, DOM provides a single source of fulfillment performance data, supporting cross-network analysis, and improvement (Norek & Gibson, 2012).

Other Execution Tools

The core supply chain execution tools are supported by a number of solutions that provide control over related resources and ancillary activities. Slotting software tools evaluate and adjust product locations based on demand velocity in the warehouse to maximize order picking productivity. Yard management systems control the flow of freight containers into and out of supply chain facilities to reduce delays time and bottlenecks. Labor management systems determine fulfillment personnel requirements based on an evaluate work volume versus engineered labor standards. And billing systems allow logistics service providers to automate their invoicing processes.

Event Management and Visibility

Each of the planning and execution software solutions contributes to the efficient and effective fulfillment of customer demand. They work well when everything goes according to plan but may falter when a disruption or failure occurs. This is particularly true for older applications. Hence, organizations need another set of tools, called supply chain event management (SCEM) applications to address this shortfall. These ancillary tools support an organization's need for accurate and timely management of information in order to maintain on-time deliveries, reduce inventory levels, and ensure that the right product is in the right place at the right time (Gourdin, 2006).

The goal of SCEM is to keep all users in the supply chain—from materials suppliers and buyers to warehouse managers and product carriers—informed about activities occurring across the supply chain. SCEM applications provide visibility of critical flows and attempts to identify any deviations from plans or schedules, as early as possible. It then triggers corrective action recommendations according to predefined rules (Otto, 2003). Using these predefined workflow rules, the SCEM system can offer suggestions that allow a decision maker to implement the best alternative in terms of cost, time, and customer service requirements.

For example, SCEM tools can monitor information regarding LCD television deliveries between an electronics factory in Taiwan and retail distribution centers in the United States. Should a container of televisions be delayed at a port or border, the stakeholders receive notification of a delay in the process. The SCEM software could recommend rerouting of the container through a different port, changing the final delivery mode, or moving the freight to an expedited carrier. The manager would use this information to make the modification or accept the late delivery.

SCEM is not limited to transportation events. It can be used to monitor the reservation of required inventory, placement of product into storage, commitment of a producer to an order, and other steps in an organization's business processes. To get started with SCEM applications, organizations must articulate the critical events in their supply chain processes, determine the types and potential severity of problems that can occur, set problem threshold levels for alerting stakeholders, and determine corrective action protocols. The goal of SCEM software is to give managers timely and clear visibility of disruptions and exceptions that have negative ripple effects on quality, cost, and customer service (Norek & Sykes, 2004).

Increasingly, software vendors are integrating SCEM and visibility solutions into supply chain execution tools such as WMS, TMS, and MES. This will increase the speed with which unexpected events that are out of tolerance with the original plan are identified and rectified (Rogers, 2011). The better these tools work together, the better the supply chain will perform. Managers will work from truth rather than assumptions and have less need to use overtime, production change orders, expedited transportation, or safety stock (Norek & Sykes, 2004).

Business Intelligence

Measuring supply chain success by KPIs is a longstanding practice. The software tools discussed throughout this chapter typically include data extraction and report generation functionality to help managers calculate performance and identify areas for process improvement. Business intelligence (BI) tools eliminate the need to manipulate data and dig through detailed reports by automating the analytical work and presenting the results in visual formats that are easier to understand (Partridge, 2013). The BI tools extract data from execution systems across the supply chain to a data warehouse where it is quickly analyzed. Fresh information is sent to frontline employees and executives for more effective planning and decision making.

In addition to the data collection and analysis capabilities, BI software supports self-service reporting, performance scorecarding versus goals, development of dashboards and other graphical report displays, and activity monitoring in support of event management. These tools can provide better access to data residing on multiple platforms without extensive information technology department support, improving the knowledge of decision makers and supporting collaboration across the supply chain. Benchmarking capabilities are also enhanced by BI tools.

The increased user-friendliness of BI software and the potential payoffs are driving adoption of the tools. When done right, BI helps the organization use root cause analysis to understand problems. In turn, stronger decisions can be made that create competitive advantage. BI opportunity areas include generating valuable insights about complex global operations, providing more granular visibility and categorization of spending, improving S&OP and demand forecasting, and resolving logistics-related bottlenecks in the supply chain (Genpact, 2013).

Tangible benefits are derived from strong BI initiatives. According to an Aberdeen Group study, leading BI software users have an average cash conversion cycle time of 39 days versus an industry average is 48 days. These top companies also achieve dramatically higher customer service levels of 96 percent compared with an industry average of 84 percent. Another performance indicator that draws a lot of attention is forecast accuracy with leaders' demand projections being accurate 79 percent of the time versus the industry average of 65 percent. Overall, these top organizations are better able to manage data and reduce costs, thanks to strong BI capabilities (Burger, 2011).

Facilitating Tools

Supply chain planning, execution, event management, and BI tools are tremendous advances over the spreadsheet tools historically used to capture and manage supply chain data. Still, the latest tools cannot be truly standalone as they require data from other sources, and managers must align their supply chain decisions to organizational goals and processes. Briefly discussed here are systems and applications that provide important facilitating links between supply chain processes, the organization, and external stakeholders. Collectively, they must work together to create an accurate, holistic view of the supply chain.

Enterprise Resource Planning

ERP systems are multimodal application software platforms that help organizations manage the important parts of their businesses. Initially concentrated on manufacturing issues, ERP systems focus on integrating information and activities across the organization (that is, the enterprise) via a common software platform and centralized database system. Key business processes linked via ERP include accounting and finance, planning, engineering, human resources, purchasing, production, inventory and materials management, order processing, and more. The centralized and shared database system ties the entire organization together, allowing information to be entered once and made available to all users. Business processes can also be automated for rapid, accurate execution (Coyle et al., 2013).

Over time, the traditional separation of supply chain technologies from ERP systems has faded. The major ERP systems vendors are developing competitive supply chain software that readily leverages the data and information already stored in the ERP system. Though these ERP vendor versions of WMS, TMS, and other tools may not be as robust as best-of-breed supply chain software, they do have the advantage of being a one-stop solution and offer common structure that reduces the installation time and effort (McCrea, 2013a).

As the ERP systems branch out to include supplier relationship management and customer relationship management (CRM), the supply chain value of ERP grows stronger. Supply chain stakeholders can access critical information and data through the ERP system to assess inventory availability, production schedules, and delivery information. This accessibility enhancement leads to improved visibility, faster and more accurate transactions, and enhanced decision making (Wisner, Keong Long, & Tan, 2011).

Supplier Relationship Management

Supplier relationship management (SRM) is a comprehensive approach to planning and managing an organization's interactions with providers of goods and services. SRM practices create a common frame of reference to enable effective communication between organizations that may have very different business practices and terminology. This is supported by SRM software that facilitates design collaboration, sourcing decisions, negotiations, and buying processes (Bozarth & Handfield, 2013). SRM software also helps the organization evaluate supplier risk, performance, and compliance throughout the life cycle of a contract.

The goal of SRM and related software is to streamline transactions, simplify information flows, and improve the procurement process (Rouse, 2005). Effective alignment of capable SRM software with strong procurement processes will yield the ability to consistently acquire needed inventory at the best available prices. Interactions will be conducted and managed in a systematic, integrated fashion across the life cycle of supplier relationships, across business units, and across functions. And, supplier assets, expertise, and capabilities will be leveraged for maximum competitive advantage (Hughes & Wadd, 2012).

Customer Relationship Management

Customer relationship management (CRM) is an information industry term for methodologies, software, and usually Internet capabilities that help an enterprise manage customer relationships in an organized way (Rouse, 2006). Organizations need to learn more about customer needs, behaviors, and demand patterns in order to develop stronger relationships with them. Though this appears to be a sales and marketing activity, supply chain managers contribute to strong customer relationships and retention through improvements in inventory availability and customer service across all channels—websites, brick-and-mortar stores, call centers, and direct sales (Wailgum, 2010).

CRM systems link up each of these points of engagement and play a critical role in coordinating and standardizing customer relationships. The collected data flows between operational systems (for example, sales and inventory systems) and analytical systems sort through transactional records for relevant information and patterns. Supply chain analysts can then comb through the data to obtain a holistic view of each customer and pinpoint areas where better services are needed (Wailgum, 2010).

Automatic Identification

By itself, supply chain software cannot provide actionable knowledge for supply chain managers. Automatic identification (auto-ID) technologies allow relevant, accurate data to be quickly captured and synchronized for use by skilled individuals in the planning and execution of supply chain processes. Barcode labels, radio-frequency identification (RFID) tags, and related hardware and software work together to recognize objects, capture information about them, and feed the data directly into supply chain systems with little to no human involvement.

Auto-ID tools facilitate total asset visibility and control of products as they move between transfer points in the supply chain. Because data can be obtained continuously, it is more up-to-date than periodically collected data such as inventory reconciliation counts. By not involving humans in the data collection process, auto-ID readings are less expensive and generally more accurate with fewer misreads and recording errors (McFarlane and Sheffi, 2003). The resulting timely, accurate data can be used to support shipment tracking and product traceability, supply chain event management, and replenishment.

Future Technology Outlook

As a supporting function for many areas, the supply chain technology landscape is ever-changing. New supply chain requirements and changing needs drive technology innovation while broad technology breakthroughs create new options for managing supply chain information more effectively. In addition, mergers and collaboration between solutions vendors drive software functionality and integration improvements. Given these opportunities, technology investment is projected to grow 61 percent over the next 5 years. Gartner forecasts that spending on supply chain software will increase to $8.9 billion in 2017 from 2012 levels of $5.5 billion (Trebilcock, 2013).

Industry observers point to three technology opportunities that could shape the way supply chain managers capture and utilize data in the coming years. Each is discussed in the following sections.

Cloud Computing

Cloud computing is emerging as a viable option for on-demand access to supply chain software applications and data sharing. The National Institute of Standards and Technology (2011) definition of cloud computing states: Cloud computing is a model for enabling convenient, on-demand network access to a shared pool of configurable computing resources (for example, networks, servers, storage, applications, and services) that can be rapidly provisioned and released with minimal management effort or service provider interaction.

The interest surrounding cloud computing is based on its economic, architectural, and strategic value. The economic value comes from the ability to pay as you go and to avoid huge up-front infrastructure investment costs. This makes supply chain software accessible to small- and mid-sized organizations. The simple, consistent environment in which application development and deployment take place is the architectural value. Finally, cloud platforms allow organizations to strategically focus on their core competencies while leaving the technical responsibilities to a third-party expert at a competitive price (Coyle et al., 2013).

Cloud computing will help organizations expand their supply chain visibility and end-to-end supply chain synchronization. By providing a control tower from which relevant supply chain information can be seen by all stakeholders, cloud computing creates one version of the truth and eliminates information time lags that hinder event management response. Problem resolution will become a routine part of the work of continuous supply-demand rebalancing (O'Marah, 2013).

Mobility Solutions

The ubiquity of smartphones, tablet computers, and other mobile devices is expected to move from the consumer markets to the workplace. The evolution of smaller, cheaper, and more mobile tools provide supply chain managers with ready access to critical decision-making information on the distribution center floor, in the factory, and offsite. Workforce visibility, cohesiveness, and communication are enhanced by mobile devices. And manual paper-based processes are often replaced by electronic processes using handheld devices that capture, track, and report data in real time (McCrea, 2013c).

The future of mobile solutions is seemingly boundless. Already, smartphone cameras can scan quick response codes for price lookup and inventory availability status, add-on devices turn tablets into inexpensive distribution data terminals, and software vendors are releasing mobile applications for smartphones and tablets (Coyle et al., 2013). In the near future, social networking capabilities for connecting with suppliers and customers, sharing documents, and creating discussion forums with colleagues and business partners is predicted to grow (Gonzalez, 2013). There is also a great deal of customer and product feedback information embedded in blogs and social media that can be analyzed and used by supply chain managers to improve product design, procurement, manufacturing, and flows (Sengupta, 2013).

Business Analytics

The upshot of all the technology used to manage supply chains is the vast amounts of data that is generated. This is both a boon and a bane for managers who must act like gold prospectors, sifting through "big data" to find the valuable nuggets of useful information. BI tools help with the historical analysis of performance that uses metrics, reports, dashboards, scorecards, and queries. They address questions related to what happened, when, and how much.

Another level of insight is needed to focus on issues of why events occurred, their likelihood of reoccurrence, and the impact of changes. Business analytics (BA) tools handle these issues through advanced quantitative analysis, data mining, and predictive modeling (Rouse, 2010). BA is also capable of analyzing large amounts of data quickly to address important supply chain issues regarding demand, uncertainty, and risks particularly for extended supply chains. Managers develop the ability to turn BA output into strong decisions that benefit the organization, supply chain stakeholders, and customers (Stock, 2013).

Chapter Summary

It has been said that information is the lifeline of business, driving effective decisions and actions. It is especially critical to supply chain managers because their direct visibility to external supply chain processes is very limited. Access to accurate and timely information provides managers with valuable cross-chain knowledge of supplier compliance, production schedules, customer demand, inventory stock levels, and transportation flows. This knowledge is essential for effective situational assessments and decisions that drive organizational success.

To facilitate these knowledge links and foster supply chain visibility, organizations are investing in information technology tools, supply chain software, and web-based capabilities. This chapter has highlighted the critical importance of information for strategic planning, tactical decision making, and process execution. Quality information must readily flow within the organization and across the supply chain to stakeholders and service providers.

Harnessing the power of information requires the use of facilitating technologies. Organizations must select tools that support optimization, speed to market, agility, collaboration, and related capabilities discussed throughout the chapter. Supply chain managers must also take steps to avoid the risks and challenges that accompany technology adoption. Most important, they must recognize that technology is an enabling tool rather than a cure-all for supply chain ills.

Selection of appropriate software that fits into a logical framework of automated decision making, exception management, cross-chain visibility, and an integrated planning and execution infrastructure will provide a strong technology foundation. The chapter presents a variety of important software solutions for planning, executing, managing, and evaluating supply chain processes, as well as emerging capabilities that will help organizations use technology to drive continuous improvement and competitive advantage.

Supply chain technology has the potential to facilitate success. To turn this potential into reality, supply chain managers must remain attuned to the growing role of information, understand the array of software options, choose solutions wisely, and overcome key implementation challenges. Doing so will help the organization gain maximum value from information technology.

References

Accenture. (2009) *High performance in a volatile world: Seven imperatives for achieving dynamic supply chains.* Retrieved January 20, 2013, from http://www.accenture.com/us-en/Pages/insight-seven-imperatives-achieving-dynamic-supply-chains-consumer-goods-summary.aspx.

ARC Advisory Group. (2012) Supply chain sophistication drives supply chain planning growth. *Supply Chain Planning.* Retrieved August 31, 2013, from http://www.arcweb.com/market-studies/pages/supply-chain-planning.aspx.

Banker, S. (2012) To understand a supply chain, you need to model it with supply chain design tools. *Logistics Viewpoints.* Retrieved August 30, 2013, from http://logisticsviewpoints.com/2012/11/12/to-understand-a-supply-chain-you-need-to-model-it-with-supply-chain-design-tools/.

Banker, S. (2013) Distributed order management: A critical application. *Logistics Viewpoints.* Retrieved September 3, 2013, from http://logisticsviewpoints.com/2013/03/25/distributed-order-management-a-critical-application/.

Blanchard, D. (2008) Top 10 demand planning strategies. *Industry Week.* Retrieved August 31, 2013, from http://www.industryweek.com/companies-amp-executives/top-10-demand-planning-strategies.

Bordner, T. (2013)The new world of global trade management. *Supply Chain Brain.* Retrieved September 3, 2013, from http://www.supplychainbrain.com/content/nc/technology-solutions/global-trade-management/single-article-page/article/the-new-world-of-global-trade-management/.

Bozarth, C. C., and Handfield, R. B. (2013) *Introduction to operations and supply chain management.* Upper Saddle River, NJ: Pearson Education.

Brady, J. (2013) The five main supply chain challenges companies face today. *The Edge.* Retrieved August 31, 2013, from: http://www.supplychainedge.com/the-edge-blog/the-five-main-supply-chain-challenges-companies-face-today/.

Burger, D. (2011) Aberdeen examines BI in the supply chain. *The Four Hundred,* 20(19). Retrieved September 3, 2013, from http://www.itjungle.com/tfh/tfh052311-story06.html.

Cecere, L. (2012) Preparing to run the race: Supply chain 2020. *Supply Chain Shaman.* Retrieved August 31, 2013, from http://www.supplychainshaman.com/uncategorized/preparing-to-run-the-race-supply-chain-2020/.

Chopra, S., and Meindl, P. (2013) *Supply chain management: Strategy, planning, and operation.* Upper Saddle River, NJ: Pearson Education.

Coyle, J. J., Langley, C. J., Novack, R. A., and Gibson, B. J. (2013) *Supply chain management: A logistics perspective.* Mason, OH: South-Western Cengage Learning.

Cudahy, G. C., George, M. O, Godfrey, G. R., and Rollman, M. J. (2012) Preparing for the unpredictable. *Outlook: The Online Journal of High-Performance Business.* Retrieved August 8, 2013, from http://www.accenture.com/us-en/outlook/Pages/outlook-journal-2012-preparing-for-unpredictable.aspx.

Enslow, B. (2006) *Global supply chain benchmark report.* Retrieved August 30, 2013, from http://www-935.ibm.com/services/us/igs/pdf/aberdeen-benchmark-report.pdf.

Ernst & Young. (2011) *Operational agility: From supply chain to integrated value chain.* Retrieved March 20, 2013, from http://www.ey.com/GL/en/Issues/Driving-growth/Growing-Beyond--Operational-agility.

Gartner. (2012) Gartner says worldwide supply chain management software market grew 12.3 percent to reach $7.7 billion in 2011. *Gartner Newsroom.* Retrieved September 1, 2013, from http://www.gartner.com/newsroom/id/2016915.

Genpact. (2013) Integrated business intelligence solutions. Retrieved September 4, 2013, from http://www.genpact.com/home/solutions/analytics-research/integrated-bi-for-scm.

Gonzalez, A. (2013) A pulse on social networking for supply chain management. *Talking Logistics.* Retrieved September 4, 2013, from http://talkinglogistics.com/2013/07/17/a-pulse-on-social-networking-for-supply-chain-management/.

Gourdin, K. (2006) *Global logistics management.* Malden, MA: Blackwell Publishing.

Hausman, W. H., Lee, H. L., Napier, G. R. F., and Thompson, A. (2009) *How enterprises and trading partners gain from global trade management.* Retrieved September 3, 2013, from http://www.scdigest.com/assets/reps/Stanford_GTM_Report.pdf.

Heaney, B. (2013) *Supply chain visibility: A critical strategy to optimize cost and service.* Retrieved August 24, 2013, from http://www.aberdeen.com/Aberdeen-Library/8509/RA-supply-chain-visibility.aspx.

Hitachi Consulting (2007) *Sales and operations planning: The basics.* Retrieved August 30, 2013, from http://www.hitachiconsulting.com/files/pdfRepository/WP_SalesOperationsPlanning.pdf.

Holcomb, M., Nightengale, T., Ross, T., and Manrodt, K. B. (2011) *20th annual trends and issues in logistic and transportation study: The new normal.* Retrieved August 28, 2013, from http://manrodt.com/pdf/Normal_2011.pdf.

Hughes, J., and Wadd, J. (2012) Getting the most out of SRM. *Supply Chain Management Review,* 16(1), 22–29.

Klappich, C. D. (2012) A strategic shift. *CSCMP's Supply Chain Quarterly,* 6(Special Issue), 43–44.

Maloney, D. (2006) More than paper savings. *DC Velocity,* 4(1), 62–64.

McCrea, B. (2012) 2012 Supply chain software users survey: Spending stabilizes. *Logistics Management,* 51(5), 38–40.

McCrea, B. (2013a) ERP vs. best-of-breed. *Logistics Management,* 52(7), 44–47.

McCrea, B. (2013b) Software vendors innovation drives growth in the transportation management systems market. *Supply Chain 247.* Retrieved September 3, 2013, from http://www.supplychain247.com/article/software_vendors_innovation_drives_growth_in_the_transportation_management/trends/D2.

McCrea, B. (2013c) Wireless & mobility: 8 trends taking us closer to visibility. *Logistics Management,* 52(8), 58–60.

McCrea, B. (2013d) 11th annual software users survey: Caution remains. *Logistics Management* 52(6): 36–40.

McFarlane, D., and Sheffi, Y. (2003) *The impact of automatic identification on supply chain operations.* Retrieved September 4, 2013, from http://web.mit.edu/sheffi/www/documents/TheImpactofAutomaticIdentificationonSCOperations.pdf.

Moore, P., Manrodt, K., and Holcomb, M. (2005) *Collaboration: enabling synchronized supply chains, year 2005 report on trends and issues in logistics and transportation.* Retrieved August 31, 2013, from http://web.utk.edu/~mholcomb/2005AnnualStudy.pdf.

Napolitano, M. (2012) Maximize your WMS. *Logistics Management.* Retrieved September 1, 2013, from: http://www.logisticsmgmt.com/view/maximize_your_wms/tms.

National Institute of Standards and Technology. (2011) The NIST definition of cloud computing, *NIST.gov*. Retrieved September 4, 2013, from http://csrc.nist.gov/publications/nistpubs/800-145/SP800-145.pdf.

Norek, C., and Gibson, B. (2012, Winter) Distributed order management: How should retailers enable fulfillment across multiple channels? *Logistics Quarterly, 2011*(12), 36.

Norek, C., and Sykes, S. (2004) Supply chain event management: Is it time to implement? *Logistics Quarterly*, 10(3). 3–4.

O'Marah, K. (2013) Get ready for supply chain in the cloud. *SCM World*. Retrieved August 29, 2013, from http://www.scmworld.com/Blog/Get-ready-for-supply-chain-in-the-cloud/.

Otto, A. (2003) Supply chain event management: Three perspectives. *International Journal of Logistics Management*, 14(2), 1–13.

Partridge, A. R. (2013) Business intelligence in the supply chain. *Inbound Logistics*, 33(4), 39–46.

Pearson Specter, S. (2013) Trends transforming supply chain infrastructure. *MHI Press Release*. Retrieved August 31, 2013, from http://www.mhi.org/media/news/12232.

Premus, R., and Sanders, N. (2005) Modeling the relationship between firm IT capability, collaboration, and performance. *Journal of Business Logistics*, 26(1), 1–23.

PwC. (2013) *Global supply chain survey 2013: Next-generation supply chains efficient, fast and tailored*. Retrieved August 28, 2013, from http://www.pwc.com/et_EE/EE/publications/assets/pub/pwc-global-supply-chain-survey-2013.pdf.

Quinn, F. (2013) Maximizing your return on investment from investment in supply chain technology. *Supply Chain 247*. Retrieved August 31, 2013, from http://www.supplychain247.com/article/maximizing_your_return_on_investment_from_supply_chain_technology/D4.

Reynolds, S., and Khan, T. (2013) *2012–2013 Transport & logistics CIO report*. Retrieved August 24, 2013, from http://www.eft.com/technologyit/it-strategy-logistics-cios.

Riley, J. (2012) ICT: What is good information? Retrieved August 30 2013, from http://www.tutor2u.net/business/ict/intro_information_qualities.htm.

Rogers, L. K. (2011) Supply chain software basics: Supply chain execution. *Modern Materials Handling*, 66(11), 26–34.

Rouse, M. (2005) Supplier Relationship Management (SRM). *SearchSAP TechTarget*. Retrieved September 3, 2013, from http://searchsap.techtarget.com/definition/supplier-relationship-management.

Rouse, M. (2006) CRM (Customer Relationship Management). *Search CRM Tech Target.* Retrieved September 4, 2013, from http://searchcrm.techtarget.com/definition/CRM.

Rouse, M. (2010) Business Analytics (BA). *SearchBusinessAnalytics, TechTarget.* Retrieved September 4, 2013, from http://searchbusinessanalytics.techtarget.com/definition/business-analytics-BA.

SC Digest Editorial Staff (2013) What is the state of supply chain planning? Retrieved August 31, 2013, from http://www.scdigest.com/assets/on_target/13-08-21-2.php?cid=7331.

Sengupta, S. (2013) 10 trends for the next 10 years. *Supply Chain Management Review,* 17(4), 34–39.

Simchi-Levi, D., Kaminsky, P., and Simchi-Levi, E. (2008) *Designing and managing the supply chain: Concepts, strategies, and case studies.* New York: McGraw-Hill/Irwin.

Stock, J. R. (2013) Supply chain management: A look back, a look ahead. *CSCMP's Supply Chain Quarterly,* 7(2), 22–26.

Trebilcock, B. (2013) Top 20 SCM software suppliers, 2013. *Modern Materials Handling,* 68(7), 32–36.

Wailgum, T. (2010) CRM definitions and solutions. *CIO.* Retrieved September 3, 2013, from http://www.cio.com/article/40295/CRM_Definition_and_Solutions?page=1&taxonomyId=3005.

Weiss, T. R. (2011) Manufacturing Execution System (MES) FAQ. *Manufacturing execution systems strategy update: Trends and tips for 2011.* Retrieved September 3, 2013, from http://viewer.media.bitpipe.com/1212985018_498/1268690867_382/Apriso_sManERP_EBook_010411_RSAebook.pdf.

Wisner, J., Keong Leong, G., and Tan, K. C. (2011) *Principles of supply chain management: A balanced approach* (3rd ed). Mason, OH: South-Western Cengage Learning.

5

MANAGING THE GLOBAL CHAIN

Global supply chains are responsible for facilitating international trade worth trillions of dollars annually. In fact, there has been dramatic growth in global trade over the last two decades as countries have shifted their attention from a predominantly domestic focus to a broader and more global view. As the level of global activity has accelerated, so too has the complexity of the supply chains that are tasked with supporting these broader and more complex trade networks.

One key element to a successful global business presence is having a superb global supply chain. The management of a global supply chain network must effectively coordinate the functional activities of the various organizations of a supply chain network (Dornier, Ernst, Fender, and Kouvelis, 1998). There are many challenges to coordinating multiple functional areas of multiple firms across a complex and geographically dispersed global network. However, if properly executed, the result can be a seamless pipeline designed to provide a continual flow of products or services to meet customer demands.

Responding to Global Customer Demand

Supply chain networks are created to support the goal of successfully fulfilling the needs of customers. Although this may sound simple enough, consider the people you know and interact with on a regular basis. Do they all desire the same products with the same features? Do they all drive the same cars and wear the same types of clothes? Clearly not, yet in many cases these are the people you have the most in common with and who may be the most similar to you.

Walk into any large clothing retailer and you will be confronted with many different styles, sizes, colors, and combinations. Walk down a grocery store aisle and examine the many different varieties of food and drink available for purchase. Companies provide a plethora of choices to you in an attempt to satisfy consumer demands and create sales revenues. Yet, while there are many choices available at your local clothing or grocery retailer, the variety of products available at your store can't possibly be all-inclusive.

Travel to a different area of the country and you will see items not available in your local stores. Travel internationally and you will discover products you never knew existed. The wide variety of products available globally is a response to the extensive diversity of consumer tastes and preferences. Yet while it would be nice to be able to acquire a customized product that perfectly meets all our individual wants and needs, total product customization for each individual consumer is costly and not practical.

One of the biggest challenges for a global company trying to efficiently serve a wide variety of customers is achieving an appropriate balance between meeting individual customer demand and designing a system that retains a sufficient level of efficiency. Process standardization facilitates efficiency. Product customization facilitates attracting new customers and retaining them. Achieving the ideal balance between these two seemingly competing initiatives is difficult. In fact, this process requires constant fine-tuning that necessitates that agility and flexibility be built into the entire global supply chain network.

A significant amount of supply chain planning emanates from the question of how to best serve the global customer base. To be able to properly serve a diverse set of global consumers, global supply chain planners must be able to devise a supply chain network capable of responding to this question. In response, most successful supply chains have attempted to develop a thorough understanding of the customers they serve. Once a sufficient understanding of customers in a particular area or region is obtained, supply chain managers can accurately segment customers by country or region. This allows supply chain planners to then begin to establish tailored service capabilities specifically designed to meet the needs of the customers of a particular country or region.

Ideally marketing managers would like to be able to offer a completely customized product ideally suited to each customer around the world. However, this must be balanced against supply chain managers who prefer to pursue a centralized strategy with as much standardization as possible so they can effectively manage the global supply chain network. Although a centralized strategy is ideal for achieving efficiency, standardized practices across the globe are not likely to fit the diverse needs associated with various markets or regions (Diederichs & Leopoldseder, 2008).

Ultimately, the consumer drives demand, which creates a continuous challenge because the tastes and preferences of consumers are not static. Couple this with the many new products arriving on the market, and it creates a situation in which consumer buying habits are in a constant state of evolution. Because of this phenomenon, today's global supply chains must be built to be adaptable to an environment of almost constant change. To operate effectively in this type of environment, supply chain managers must be able to identify supply chain trends and to alter supply networks as necessary. Supply chain managers who can do the best job of spotting and adapting to the latest trends and adjusting their networks accordingly can meet their objective of supporting businesses to meet consumer demand (Lee, 2004).

If this objective is to be consistently met, adaptability is key. From the final consumer all the way back through the supply chain to the raw materials suppliers, a complex network of entities must effectively perform a wide variety of functions to create a seamless acquisition, manufacturing, and delivery system. The challenges of managing this type of system are heightened when you consider the many varieties of products, types of suppliers, locations of manufacturers, and varieties of distribution options available to achieve the goal of serving a diverse set of customers. Supply chain complexity, especially on a global scale, creates monumental challenges for any supply chain manager.

Many of today's successful businesses have a complex and global network of partners. Some of these partners may contribute to the effective supply of raw materials, component parts, and other items or services necessary for success. Other partners may assist an organization in achieving their manufacturing or product assembly and/or distribution goals. Still other partners may focus on developing new markets; enhancing sales opportunities; or building and/or expanding a broad, global customer base. Regardless of the specific role of the supply chain network partner, significant challenges arise when conducting business globally.

Financial and Investment Related Factors

Doing business internationally can bring unique financial challenges and complicated investment decisions that go well beyond those of the typical domestic business venture. There are a myriad of seemingly endless scenarios that confront a business contemplating global expansion in order to take advantage of a new market opportunity. How to enter the market? Perhaps a joint venture with a foreign business partner is attractive. Perhaps some type of licensing or franchising agreement is more appropriate considering the circumstances. What are the potential risks and rewards associated with entering the new market? What are the advantages and disadvantages associated with each entry option? Although it is unnecessary to cover every possible global business decision in this forum, it is important to note some of the most common issues relevant to most global business practitioners.

One common issue most global businesses face is the impact of having to deal with multiple and different currencies. A U.S.-based company that conducts domestic business operations deals only with the U.S. dollar. This makes pricing and payment practices relatively easy to perform. However, what happens when the same U.S.-based company ventures overseas to begin using a European supplier? That supplier might want to conduct business in Euros, its standard or preferred currency for these types of international transactions.

Consider the following scenario. The U.S.-based company enters into a contract with the European supplier to provide 10,000 finished products at a cost of 10 Euros each. The U.S.-based company plans to bring the finished product to the United States and sell them to its American customers for $20 each. At the time the agreement was signed, 1 U.S. dollar was worth .8 Euros. Therefore, the U.S. buyer anticipates owing the European supplier 100,000 Euros, which at the time of the agreement would be equal to $125,000 (U.S.) (10,000 units * 10 Euros /.8 currency ratio of $ to Euro). The U.S.-based buyer incurs an additional $25,000 (U.S.) for the logistics and packaging associated with getting the items ready and in place to sell to U.S. customers.

The U.S.-based company anticipates sales revenues of $200,000 (10,000 units * $20 U.S.) once the finished products are all sold to its U.S. customers. With sales revenues of $200,000 and anticipated costs of only $150,000 ($125,000 for the products and $25,000 for logistics), the company figures on a tidy little profit from the venture. When the invoice arrives from the European supplier, predictably it is for 100,000 Euro. Unfortunately, when the U.S.-based buyer goes to the bank to arrange for payment, it is told that it will take almost $143,000 (10,000 units * 10 Euro / .7 currency ratio of $ to Euro) to pay the invoice! What happened? Very simply, currencies fluctuate in value relative to other currencies.

Between the date the agreement was signed and the date the invoice was due, the U.S. dollar lost value relative to the Euro. Although the U.S. dollar was worth .8 Euros when the original agreement was signed, unfortunately by the time the invoice was due, the U.S. dollar was only worth .7 Euros. Therefore, it took more U.S. dollars than previously anticipated to cover the 100,000-Euro invoice. Now instead of a $50,000 profit, the firm is looking at a smaller profit of about $32,000. Although there are ways to minimize your risk exposure to currency fluctuations, this is merely one example of how conducting global business operations creates unique challenges beyond those encountered in a typical domestic business venture.

Cultural and Communication Factors

Companies doing business with entities located in foreign countries throughout the world are also likely to face numerous cultural challenges. Cultures throughout the world are very different, which can create peril for anyone venturing into unfamiliar territory. Cultural norms can affect many different issues and have a potentially dramatic impact on business practices. For example, something as simple as a cultural difference in the way a delivery date is communicated can cause mass chaos throughout the supply chain.

Consider the following example. A U.S.-based company elects to source a component part for its automotive manufacturing plant from an overseas supplier. The U.S. buyer clearly states in the contract that it needs to receive delivery of the component part on or before 02-03-2014. Although most Americans interpret this to be February 3, 2014, in other parts of the world 02-03-2014 is interpreted as the second day of the third month of 2014! Unfortunately this is March 2, 2014, roughly a month after the part is needed by the U.S. manufacturer! Consider the chaos created when the U.S.-based automotive plant does not receive its component part order on February 3rd because the foreign supplier believes the parts are not due to their customer until March 2nd. This is an example of how something as simple as a different cultural convention can unknowingly disrupt a global supply chain and have costly consequences.

Similar to the diverse cultural norms that are encountered throughout the world, numerous languages and various dialects of different languages are spoken across the globe. This variety can lead to a lack of ability to practice effective written or oral communication skills. Translation from one language to another may not be exact, which can create confusion or even alter the meaning of a conversation or written communication such as a contract. Although there are ways to address dealing with a foreign-speaking business partner, it is clear that global business operations often involve challenges and complexities beyond those of the typical domestic counterparts. These challenges can be overcome, but they must be anticipated and planned for in advance of entering into an agreement with global partner.

Understanding cultural norms is also vitally important when assessing new market opportunities. Without a sufficient understanding of the people who make up the potentially new marketplace, it is nearly impossible to capture the attention of consumers and convert them to loyal customers. History is full of new product introductions to unfamiliar markets that, for whatever reason, failed miserably. Although there are a variety of specific reasons for these failures, most have one overarching element in common: an insufficient understanding of the marketplace.

Legal, Political, and Economic Environmental Factors

Although legal and political environments rarely provide 100 percent guaranteed clarity, a U.S.-based firm conducting domestic operations usually has a comparatively better understanding of the legal and political environments in which they operate on a daily basis. Companies with global operations that touch many different areas of the globe are faced with a diverse set of political environments and many unfamiliar legal environments. Both the political and legal environments can have a profound impact on the decisions made by global business practitioners and must be considered carefully as part of the formulation of a global business strategy.

Although there are many important issues that fall under this broad category of factors, one common element requiring the attention of global business operators is the impact of tariffs, taxes, trade quotas, and other potential trade impediments on their global business strategies. As countries and regions have started to discover that attracting businesses can have a positive impact on their economic situation, they have taken steps to attract global business operators. One popular way to help accomplish this is to develop and participate in some type of trade agreement that is designed to facilitate global trade.

This is accomplished through the use of agreements between two or more countries or areas of the world in which each participant agrees to lift most or all tariffs, quotas, special fees and taxes, and other barriers to facilitate trade between the entities who are part of the agreement. These types of agreements intend to allow for faster and more efficient trade, enhancing business between the countries or areas involved in the agreement. Ideally, these types of agreements should benefit all the participants of the agreement. As these free trade agreements have grown in popularity, global trade between various countries of the world has grown fairly dramatically, leading to a greater number of global supply chains.

Global Business Channels

Global logistics must support the global supply chain. Effectively supporting the global supply chain involves far more than simply executing the physical flow of goods from their origin to destination. Without the effective flow of information between the exporter(s) and importer(s), the product will not move smoothly through the global supply chain. Furthermore, the flow of payment(s) is also critical to the timely completion of the successful international transaction. The three intertwined and critical functions depicted in Figure 5-1 are typically referred to as the transaction, communication, and distribution channels (Wood, Barone, Murphy, & Wardlow, 2002). Without all three channels operating effectively in a coordinated manner, goods moving through the global supply chain network are likely to encounter significant problems and delays.

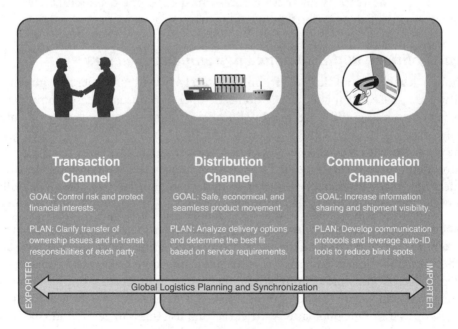

Figure 5-1   Global channels

Transaction Channel

When purchasing goods, paying for them, and preparing for their movement, the buyer (importer) must take steps to minimize its risk exposure and protect its financial interests. The buyer who is importing the goods should negotiate specific details with the seller (exporter) that go well beyond the basics of product quality, price, and quantity. There are many other critical issues that need to be addressed in the contractual agreement between a buyer and a seller before finalizing a deal. For example, in global transactions, it is vitally important to agree upon the location and point in time at which legal title for the goods transfers from the exporter to the importer.

Agreeing on the exact location of transfer of ownership is critical to understanding the potential risk of loss to the company and any potential liability that may be associated with the transaction. Once the specifics of the transaction are defined, it is easier to make the correct decisions about the mode(s) of transportation to be used, the carrier(s) to be selected, the level of insurance coverage required, and the routing of the product from its origin to its destination. Transfer of ownership also determines who is responsible for the payment of transportation services, insurance costs, and other expenses such as import duties. In addition, ownership helps to determine who holds the responsibility for compliance with government regulations; management of the goods while in transit; and the financial liability in the event of freight damage, loss, or a delay in delivery.

The transaction channel is also responsible for tasks related to payment for the goods. In a global transaction, both the exporter and importer tend to be at greater risk than in a domestic transaction. The exporter is typically concerned about the risk of nonpayment by the importer. As a result, the exporter may want to receive payment in advance of product shipment. However, typically the importer would rather receive and inspect the products prior to initiating payment. This allows the importer to obtain shipment delivery and inspect the product to determine that it meets all the agreed-upon criteria such as quality, quantity, and other specifications. As a result, the buyer would rather wait to initiate payment until after delivery. There are several payment process options available that can moderate the risks faced by both parties when dealing with these types of payment issues.

Communication Channel

Clearly the logistics networks that service today's global supply chains are complex and produce many managerial challenges. In response, many entities are utilizing technology to enhance supply chain-wide visibility to help establish more efficient logistics processes. One of the biggest challenges associated with a global supply chain is the capability to track products, maintain shipment visibility, and control freight as it moves across the borders of various countries through its long and potentially tumultuous journey.

During this process, the product may be handed off between carriers and intermediaries multiple times as it navigates its journey across the globe to its destination.

During this process, the timely sharing of information, aided by the use of technology, can vastly improve shipment visibility and, in some cases, allow for shipments to be redirected or redeployed during shipment. The use of technology to enhance information about the global shipment helps to achieve the proper freight documentation. Correct and complete documentation of freight helps to ensure compliance with government regulations and facilitates the uninterrupted flow of goods through potential bottlenecks at border crossings and ports. Conversely, without the proper and complete documentation to accompany a shipment, product delivery delays are likely to be the result.

In today's world of advanced technology, one might believe that paperwork requirements should be simple to comply with. In reality, many significant communication channel challenges remain. A tremendous number of documents are involved in a global transaction, especially when compared with a domestic transaction. In addition to the number of documents required, the types of documents may vary by the country of export, country of import, transportation companies providing services, banks facilitating payment, and even importer and exporter. Figure 5-2 depicts the primary categories of documentation.

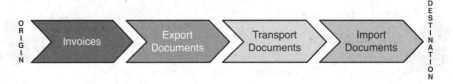

Figure 5-2   Global documentation requirements

Many critical documents are not yet in electronic format, and copies must be distributed to each party involved in the transaction. To help surmount these challenges, some organizations have turned to global trade management (GTM) systems for help. These solutions avoid paper-based, manual intervention processes, help organizations reap the benefits of free trade agreements, and achieve resiliency (Bordner, 2013).

These systems are highly automated software systems that connect cross-border trading partners electronically to provide stronger import and export processes, global logistics execution, and trade agreement control. GTM benefits include more effective collaboration and messaging, lower in-transit inventory levels, lower expediting expenses, reduced broker fees, reduced cargo insurance costs, and a variety of financing benefits. In fact, exporters gain benefits equivalent to 1.7 percent to 2.4 percent of net sales, whereas importers gain benefits equal to 0.6 percent to 2.2 percent of net sales (Hausman, Lee, Napier, & Thompson, 2009). However, many countries do not have an infrastructure that is sufficient to support the effective handling of electronic documentation, hampering the effectiveness of GTM systems. Couple this with the many different documents required to successfully complete a global transaction, and it becomes clear that many communications-related challenges remain for global supply chain managers.

Distribution Channel

Managing an extended transportation network involving significant distance, considerable complexity, and a number of participants increases the potential for disruptions. Couple this with the fact that freight may travel through multiple facilities and multiple countries, and be touched by numerous intermediaries and carriers, and effective management of product movement becomes a tricky proposition for many global supply chain managers. Furthermore, transportation infrastructure, regulations, and service options vary from country to country. As a result, global freight faces a much greater level of risk of erratic and extended transit times, freight stoppages, visibility problems, and an overall loss of control when compared with the typical domestic freight transportation network.

Although many global supply chain managers may want to raise the white flag of surrender, they must instead continually monitor and actively manage the distribution channel. A key part of this process is to recognize and act upon the need for effective transportation planning when addressing variables such as the mode(s) to be used, carrier(s) to be selected, and route(s) to be used. It is necessary to properly match freight to the most appropriate mode to facilitate the safe and cost-efficient movement of goods through the supply chain. Conducting a thorough and complete carrier selection process increases the chances of a reputable transportation service provider being selected. These suppliers should have significant levels of experience in the key markets to be served, extensive capabilities and appropriate equipment availability, and a strong customer orientation. In addition, selecting the optimal route from among the wide variety of options can reduce risk levels by providing for greater freight protection and more consistent service levels.

Channel Interactions

Global supply chain managers are faced with having to address extended distances, longer transit times, and multiple carriers involved in servicing the global supply chain network. Clearly, careful consideration of the key aspects of all three types of channels discussed previously is essential to success. When focused on the transaction channel, managers must focus on a wide variety of issues such as transfer of ownership, control of freight, and the effective execution of payment. As managers transition to focusing on the communication channel, they must understand and comply with documentation requirements, be aware of the need to interact with multiple government(s) entities, and manage a diverse and sometimes unfamiliar group of sellers and transportation/logistics service providers. Managers must also continually evaluate the distribution channel by focusing on the mode and carrier selection process, and continue to address important routing and scheduling issues. Perhaps the trickiest and most important part of the process is to work toward the effective coordination of all three channels in a manner that facilitates the smooth and seamless movement of products from their origin to destination.

Global Supply Chain Network Functions

As global supply chain networks have grown in popularity, it has become increasingly evident that numerous functional areas of a global supply chain must come together to support an effective global business operation. The most commonly recognized and discussed broad functional areas include supply management/purchasing, manufacturing/assembly, and logistics/distribution. Each of these key areas is discussed in the following sections.

Supply Management/Purchasing

If you consider the consumer to be at one end of the supply chain, at the other end you are likely to find a raw materials supplier. Figure 5-3 indicates that raw materials must first be acquired and then sent through a transformation process that ultimately produces a product that is useful to the consumer. The acquisition portion of this process is often referred to as *purchasing* or *supply management*.

Figure 5-3   Supply chain transformation process

The supply management function derives much of its importance to an organization by providing a continuity of supply in a manner that contributes to cost efficiency and enhances operational effectiveness. As supply managers try to maximize the contribution of their functional area to the organization, they often look at global sourcing opportunities. If executed properly, global sourcing can create tremendous opportunities for an organization, and it is a primary reason for the rapid rise in use of a diverse and global set of suppliers. Although there are numerous potential advantages associated with pursuing this type of strategy, sourcing managers are also discovering the many challenges that are often created with a global sourcing program.

Consider an American electronics manufacturing company that is pursuing a global sourcing strategy to acquire a critical component part for its televisions. The television manufacturer has identified a high-quality electronics component supplier in Southeast

Asia that can significantly reduce the cost per unit of acquiring a critical component. However, the cost-savings achieved by using the overseas supplier needs to be weighed against other issues created by pursuing this type of strategy.

Although using the electronics manufacturer may appear to save money by reducing the cost per unit for a critical component part, there might be a significant increase in the logistics costs incurred to deliver the item to the manufacturing facility. Although the cost per unit may be lower, the overall "landed" cost of the item may actually increase. In addition to the acquisition cost, this type of global sourcing practice can have an impact on many other areas of the business.

Operational effectiveness can be affected by this type of global sourcing strategy (Chopra & Meindl, 2010). In the electronics manufacturing example, the foreign supplier is located halfway around the globe from a buyer's manufacturing facility. Distance alone can significantly increase the risk of a costly supply disruption if a predetermined delivery schedule cannot be met. Perhaps the product arrives on time, but a costly manufacturing disruption is created because of a quality concern. The quality issue must be resolved, delaying the manufacturing process and negatively affecting the buyer's production schedule.

Other characteristics may also affect the decision of which foreign supplier to use. For example, if two different suppliers are offering a similar cost per unit, but one supplier can consistently product a far superior average transit time, the actual landed cost may be quite different. The supplier with the longer average transit time may require that additional inventory be carried to help guard against costly stockouts. This extra inventory may be costly, which could create a significant difference in the landed cost of an item, depending on the supplier selected. Variation in transit time is another potentially important factor. If one cannot rely on a supplier to deliver when promised, it will be necessary to carry safety stock if the delivery does not arrive as promised. This safety stock equates to holding additional inventory at additional costs.

Supply managers with good skills and strong relationships with a variety of current and potential suppliers can balance many of these issues. In these cases, the supply management function can create a competitive advantage for the company and save the organization large sums of money in the process. Although establishing an effective and reliable sourcing program is critical to global business success, unforeseen events can also occur. When a costly disruption occurs, it can create a tremendous amount of havoc that ripples throughout the entire supply chain network. To help avoid the negative consequences of any type of disruption, a global supply chain needs to develop contingency plans that are available for implementation in a timely manner when necessary.

Continuing the previous example of an American electronics manufacturer sourcing globally, if it is reliant on a global supply chain network, it should be astute enough to construct a resilient supply chain that includes multiple redundancies built into the network. Ideally, the resilient supply chain will help to avoid a disruption from occurring. However, if a disruption does occur, having redundancies built into the supply chain allow for a rapid recovery and should help to minimize the impact of the disruption (Simchi-Levi, Kaminsky, & Simchi-Levi, 2008).

Organizations further compound the logistics challenges created by global sourcing by also trying to create a lean supply chain. In an attempt to reduce costs and remain competitive, supply chain managers tend to limit safety stock by reducing on-hand inventories. This further adds to the importance of having a resilient and agile supply chain network.

Managers anticipate perfect deliveries that are void of any interruptions. In the event of a disruption, they expect end-to-end supply chain visibility and a resilient network that can respond rapidly when an unexpected issue arises. Implementing these types of lean supply chain practices can further elevate the risk associated with a disruption and increase the recovery costs associated with a global supply chain disruption.

In today's global supply chain environment, products often move thousands of miles through a number of countries and are handled by numerous people representing different entities (Coyle et al., 2013). A byproduct of this type of system is that, in spite of the best efforts to maintain product velocity throughout the supply chain, items periodically sit still. Material handling throughout the supply chain network must properly move the product through the transportation, storage, processing, customs, or manufacturing facility at the appropriate time. Certain raw materials, component parts, or other items must be available when needed at the correct production station, loading dock, and transportation terminal to successfully navigate the supply chain network as designed. The right product must be in the right facility at the right place, in the right condition, and at the right time to avoid a costly supply chain disruption.

Today's global supply chains are designed to prevent inventory from standing still. Inventory is costly and, although inventory shows up on a firm's balance sheet as an asset, many supply chain professionals argue that inventory can quickly become a liability. Companies desire to achieve a high velocity of inventory that moves rapidly through the supply chain. Supply chain disruptions can be costly for a variety of reasons, including the increased risk of stockouts, delayed customer delivery, increased risk of loss and damage, and the potential for a ripple effect of delays and shutdowns throughout the entire network.

Current trends suggest that top management is likely to demand more of the supply management in the future. After all, in many organizations, the supply management function is responsible for a large portion of the total costs incurred by a company. Therefore, no longer will achieving a low per-unit cost and avoiding the largest of problems be enough to define success in the organization. In many cases, the supply management function is now expected to ensure a continuity of high-quality supplies in a manner that reduces costs while enhancing profitability through improved operational effectiveness. With globalization, the role of supply management is changing in many organizations, and its overall importance to achieving key corporate strategies is clearly increasing.

Manufacturing/Assembly

Manufacturing and assembly processes are vital to the transformation process that takes place as materials and component parts transition to become useful products for consumers. As with most functional areas of a supply chain, manufacturing and assembly processes vary widely depending on the goal(s) being pursued. One relatively common decision that confronts many of today's global business entities is whether to manufacture or assemble the product internally or outsource the responsibility to an external entity.

As globalization has increased in popularity, so too has the use of contract manufacturing companies. Many organizations now elect to identify a high-quality contract manufacturer and outsource the manufacturing or assembly of their products to an external entity. Although this strategy can certainly be appropriate in many instances, the decision whether to make the product or enter into a contract to buy the product from an external entity requires a thorough examination of a myriad of issues.

One issue of paramount importance is the evaluation of production cost-savings relative to the risk and potential impact of a supply chain disruption. In order to fully evaluate this issue, it is important to understand the manufacturing environment. Often, manufacturing and inventory management models are classified as being either a push-based or pull-based model.

A push model schedules orders for production or orders goods in advance of customer demand based on a forecast. Manufacturers push the finished product through the supply chain to the final consumer. A pull model is based on building the goods to meet customer demand once it is known. The product is pulled through the supply chain by the order created from the customer. Many desire to use a pull-based model to minimize waste and reduce inventory throughout the supply chain.

Pull-based systems are common depending on the circumstances surrounding the product. For example, consider a customer who decides it is time to repaint a room in their house. They enter the paint store and are greeted with literally thousands of color choices! How can each paint store possibly stock each of the many color options available to the customer? Simple; they don't. They wait until the customer selects the color of paint they desire. The paint store then takes a can of white paint and adds the appropriate amount of dye to create the color the customer desires.

This pull-based approach drastically cuts down on waste and allows the paint supplier to customize the product to the exact desires of the customer once they are known. Can you imagine the waste associated with trying to forecast how many gallons of each color of paint to stock? Using the pull-based system prevents the paint store from having to forecast demand for specific colors of paint. Instead, it simply has to forecast the amount of base color paint to stock. This is a much simpler proposition.

Push-based systems tend to rely on a forecast of customer demand. This approach can be difficult. Can you imagine how difficult it would be to accurately predict demand for each item sold by a company! Unfortunately, some products do not lend themselves to a pull-based system as well as the preceding paint store example. It is more difficult to walk into an appliance or electronics store, or automotive dealership, and customize a product to your desires and receive immediate delivery. In these cases, the supply chain must, at a minimum, use some of the elements of a push-based manufacturing or inventory system.

Many consumers desire to walk into a store, purchase what they want, and walk out with the product in hand. Although appliance, electronics, and automotive manufacturers all have the capability to customize a product to your exact specifications, there are lead time and expense considerations associated with these types of product customizations. As a result, many of these manufacturers offer a variety of products with different features that should appeal to a broad set of consumers. They don't customize products; instead, they segment customers and then offer a limited number of products they believe will appeal to a particular segment of customers (Keegan, 2013). This allows the manufacturing or assembly operation to retain a sufficient level of operational efficiency while retaining the ability to satisfy the desires of a broad range of potential consumers.

When selecting potential contract manufacturing companies, there are a number of variables that must be evaluated before a competent decision can be made. Transit time, delivery reliability, product quality, product consistency, and damage rates are just a few of the variables that must be evaluated in light of the specific circumstances being faced. In some cases, a delivery delay may not be critical, whereas in other situations, a delivery failure could be catastrophic.

Once goods are manufactured or assembled, they require protection as they move through the supply chain. Packaging is always important, but particularly critical when products are traveling long distances through a global supply chain network. Packaging can serve multiple purposes. Not only can packaging help to provide product protection but it can also serve to promote goods and convey valuable information to consumers. When product protection packaging is the focus, the production facility is likely to take an active interest in the process because it is likely to end up being responsible for packaging the goods for shipment. As a result, the package size, shape, and material used can all greatly affect the ability to maintain product integrity and conduct packaging duties in an efficient manner. Although packaging is not as costly as transportation, typically a significant portion of logistics costs can be attributed to packaging.

Packaging not only affects marketing and production but also many other logistics activities. The size, shape, and type of packaging material influences the type and amount of material handling equipment required and can directly affect how goods are transported or stored. The easier it is to handle the product, typically the less costly the logistics costs. Different products and situations call for different packaging requirements, so shippers should work closely with carriers to find packaging methods that will meet everyone's needs.

Distribution/Logistics

Suppliers, manufacturing facilities, distribution centers, and customers are often dispersed across multiple countries or continents. This makes the logistics associated with servicing today's increasingly complex global supply chain networks a challenging and critically important component of the network. When placed on this type of global scale, integrated logistics activities can display different characteristics and result in dramatically increased complexity when compared to their domestic counterparts.

The overall importance of the relationship between global trade and the logistics function is highlighted in a recent Stanford University research effort. Researchers (Hausman et al., 2009) found that the cross-border trade processes can consist of up to 106 steps, many of which directly focus on logistics activities such as transportation, distribution, warehouse management, packaging, inventory management, material handling, and information systems. Documentation requirements can vary depending on a wide variety of factors. Some of the most common factors include the country of origin, destination country, transportation route and mode selected to move the product, and type of product(s) being shipped.

Although all the key logistics functions must be performed in a high-quality and coordinated manner to seamlessly support the global supply chain, often the most important and costly logistics function is transportation. Without effective transportation, it is difficult if not impossible to execute a well-designed global supply chain network. Transportation does more than just link networks together. In fact, an effective transportation system forms the backbone of a sound economy and plays a key role in both domestic and international economic growth.

There are five major modes of freight transportation: airlines, motor carriers, pipelines, railroads, and water carriers. Each of these modes has distinct characteristics that give it advantages over the others. The most appropriate mode of transportation depends on the situation being faced, as illustrated in Table 5-1.

Table 5-1   Modal Capabilities

Mode	Strengths	Limitations	Primary Global Role	Primary Product Characteristics
Water	High capacity Low cost Carrier availability	Slow Door-to-door accessibility	Highly efficient, long-distance transfer of containerized finished goods, bulk materials, and equipment.	Low value Raw materials Bulk commodities Containerized finished goods
Air	Speed Freight protection Flexibility	Door-to-door accessibility High cost Low capacity	Fast movement of high value goods and urgent shipments.	High value Finished goods Low volume Time sensitive
Truck	Accessibility Fast and versatile Customer service	Limited capacity High cost	Origin-port and port-destination intermodal container movement. Trans-border flow of goods between adjacent countries.	High value Finished goods Low volume
Rail	High capacity Low cost	Accessibility Inconsistent service Damage rates	Move large shipments of bulk materials between countries. Longer distance port-destination intermodal container movement.	Low value Raw materials High volume
Pipeline	In-transit storage Efficiency Low cost	Slow Limited international network	Transfer large volumes of critical fuel products and chemicals between countries.	Low value Liquid commodities Not time-sensitive

Some of the most common variables affecting mode selection include accessibility, capacity, transit time, reliability, safety, and cost (Novack, Gibson, and Bardi, 2011). These variables are highlighted in Figure 5-4 and are briefly discussed in the following sections.

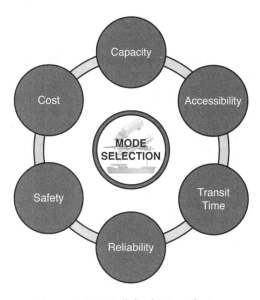

Figure 5-4   Modal selection factors

Accessibility

Transportation managers must consider the capability of each mode to reach origin and destination facilities and provide service over the specified route. A number of variables can affect a mode's accessibility including the geographic limits of a mode's infrastructure or network and the operating scope authorized by governmental regulatory agencies. Each global supply chain network may have unique requirements that help dictate the practicality of using a particular mode of transportation.

Capacity

The amount, weight, or form of a product being transported can render a mode infeasible or impractical. Transportation managers must match the capacity of a mode to the size and nature of the product being moved. Some modes are well-suited to handling a large volume of goods in an economical fashion; others are better suited to smaller goods and shipments requiring rapid movement.

Transit Time

Time is a key consideration in mode selection. Transportation affects inventory availability, stockout costs, and customer satisfaction. Transit time is the total elapsed time that it takes to move goods from its origin to its destination. Note that this is not just actual transit time; it is travel time from origin to destination. It can include time for pickup activities, terminal handling, the actual line haul movement, and customer delivery. Overall modal speed and capability to handle pickup and delivery responsibilities greatly affect actual total transit time.

Reliability

Transit time reliability is often considered to be more important than speed when determining the appropriate mode to select. Reliability refers to the consistency of the transit time provided by a transportation mode. Reliability affects predictability and makes it easier to forecast inventory needs, schedule production, and determine safety stock levels. Internationally, reliability is affected by, among other things, issues such as distance of travel, port congestion, weather, security requirements, border crossings, customs practices, and other variables that can create an unwanted supply chain disruption.

Safety

Goods must arrive at their destination in the same condition they were in when they were tendered for shipment at their origin. Proper precautions must be taken to choose a mode with the capability to protect freight from damage due to poor freight-handling techniques, inferior ride quality, and accidents. Fragile products may need to be shipped via modes with the best ride quality; bulk products tend to be more difficult to damage in transit.

Cost

Transportation cost is almost always an important consideration in the modal selection decision. Transportation costs include the rate charged for moving freight from origin to destination, plus any accessorial and terminal fees for additional services provided. Product value must be factored into the cost analysis. For example, if a company spends too much on transportation relative to the value of a product, it cannot sell the product at a competitive price. Thus, water, rail, and pipeline are generally more suitable for low-value commodities; truck and air costs can be more readily absorbed by higher-value finished goods. However, when operating in a complex and global supply chain network, not all the modes may be viable options. As a result, considerable care needs to be taken when undertaking the modal decision-making process.

Although the preceding factors may be the most common factors, other factors can also affect mode selection. The nature of a product—size, durability, and value—may eliminate some modes from consideration because they cannot physically, legally, and safely handle the goods. For example, a global supply chain manager faced with shipping a heavy and dense bulk material may be limited to water carriage while a manager faced with a short overseas delivery window may be forced to use an air carrier. Although in this example there may be only one viable option, in many cases supply chain managers must examine several alternatives in the hope that an ideal outcome can be identified.

For example, consider a global retailer that currently uses ocean carriage to ship goods from an overseas production facility to its U.S. retail distribution network. The justification for using ocean shipping is to save on transportation costs when compared with the alternative mode of air carriage. Although it is likely to be true that ocean shipping rates are lower than air carriage rates, the global supply chain manager must evaluate this decision in the context of the impact on the entire supply chain.

Air carriage, although typically more expensive, is also much faster than water carriage. As a result, there is less of a need for inventory. Furthermore, customer service levels may be positively affected due to the quicker transit time that may result in an increase in responsiveness to the customer. In fact, although transportation costs for air carriage may increase, total supply chain costs may actually decrease when you include the impact of cost-savings on other functional areas such as inventory.

Shipment characteristics such as size, route, and required speed are all important considerations. Modal capacities must be matched to the total weight and dimensions of shipments, whereas modal capabilities must be matched to customer service requirements. Combined, these considerations tend to limit modal selection to two or three realistic options, one of which may be intermodal transportation. As illustrated in Figure 5-5, intermodal transportation uses multiple modes of transportation to complete the move from an origin to a destination. Intermodal transportation is relatively common when conducting global business operations.

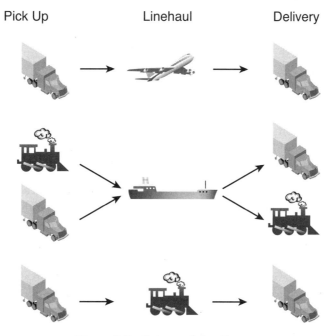

Figure 5-5 Intermodal options

The general strategy regarding modal selection focuses on determining which mode or combination of modes best suits the requirements of the global supply chain network. This long-range decision requires an analysis of the best fit and balance between modal capabilities, product characteristics, supply chain requirements for speed and service, and transportation cost. Short of major changes in price, infrastructure, service quality, or technological capabilities of a mode, this decision does not need to be revisited frequently.

Carrier Selection

Once the modal decision is complete, the next step in the process is focused on the selection of individual service providers within the chosen transportation mode. Much like the modal selection process, the carrier selection process is based on a variety of shipment criteria and carrier capabilities such as geographic coverage, transit time average and reliability, equipment availability and capacity, product protection, and freight rates.

One major difference between the modal and carrier selection processes is the number of options available. Although there are a limited number of modal options available, a comparatively broader array of carrier options exists. Typically there are numerous motor carrier options available and many of them cross country borders and offer international transportation services. Depending on the service region, there can be several major railroads that serve many global origin and destination combinations. Often there are multiple ocean and air carriers that serve each of the major intercontinental trade routes. However, the actual number of viable carrier options may vary significantly depending on a number of critical factors such as on-time delivery and pick-up, technical capabilities, carrier response to irregularities, information-sharing capabilities, freight damage rates, financial stability, dependability, and total transit time.

Although the modal decision is revisited relatively infrequently, the carrier selection process often requires more active and frequent attention. Carriers may be changed periodically. However, most of the attention is on monitoring performance levels of previously selected carriers and working with them to maintain or enhance the relationship.

Often the selection of the appropriate carrier(s) is focused on contracting with a limited number of carriers, as highlighted in the third step of Figure 5-6. This strategy helps an organization to leverage their purchasing dollars and allows them to become a more important customer of the carrier. This relationship-based approach allows the carrier to gain a better understanding of freight flows and requirements, and allows the organization to effectively monitor carrier performance levels. Even with strong alliances with a few key carriers, transportation managers serving a global supply chain must remain vigilant about monitoring carrier performance, rates, and financial stability. They must also have a backup plan to protect the continuity of product movement in the event of a disruption in carrier service.

When logistics managers properly address these activities, lead times and operational costs are kept under control. However, if these activities are not managed effectively in a global supply chain, logistics errors in cross-border trade situations create unnecessary costs, penalties and fines, delays, customer dissatisfaction, and lost revenue. For these reasons, it is vitally important to understand some of the common trading terms recognized globally.

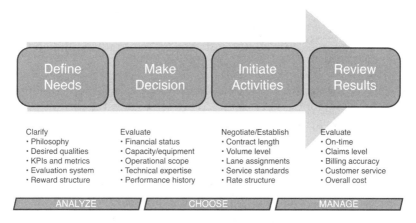

Figure 5-6   Carrier selection process

Global Trading Terms

The International Chamber of Commerce has established a harmonized set of selling terms designed to reduce the complexity and potential for confusion involving international shipments. INCOTERMS (INternational COmmercial TERMS) are a set of three-letter standard trade terms most commonly used in international contracts for the sale of goods. These INCOTERMS provide internationally accepted definitions and rules of interpretation for most common commercial terms. INCOTERMS 2010 became effective on January 1, 2011. This is the most recent revision of INCOTERMS that presented 11 specific INCOTERMS available to exporters and importers to use.

INCOTERMS are designed to inform the sales contract by defining the respective obligations, costs, and risks involved in the delivery of goods from the seller to the buyer. Although INCOTERMS are useful, they do not constitute a contract between the buyer and the seller of goods. They also do not supersede the law governing the contract, define where title transfers or address the price payable, type of currency to be used for the transaction, or specify credit terms. These items must all be specifically defined in the terms of the sales contract and by the governing law.

INCOTERMS are grouped into two classes. The majority of the terms (7 of the 11) apply to any mode of transportation; the remaining four terms apply specifically to maritime transportation. INCOTERMS are typically expressed as three-letter acronyms with a named location and the INCOTERMS version used to avoid any confusion.

Terms Available for Any Transportation Mode

A number of abbreviated terms are commonly used to describe transportation modes and situations. Among the most common are the following:

- **EXW**—EX Works (... named place of delivery). The Seller's (exporter's) only responsibility is to make the goods available at the Seller's (exporter's) premises. The Buyer (importer) bears full costs and risks of moving the goods from the Seller's (exporter's) location to the destination.

- **FCA**—Free CArrier (... named place of delivery). The Seller (exporter) delivers the goods, cleared for export, to the carrier selected by the Buyer (importer). The Seller (exporter) loads the goods if the carrier pickup is at the Seller's (exporter's) premises. From that point, the Buyer (importer) bears the costs and risks of moving the goods to their destination.

- **CPT**—Carriage Paid To (... named place of destination). The Seller pays for moving the goods to their destination. From the time the goods are transferred to the first carrier, the Buyer bears the risks of loss or damage.

- **CIP**—Carriage and Insurance Paid to (... named place of destination). The Seller pays for moving the goods to their destination. From the time the goods are transferred to the first carrier, the Buyer bears the risks of loss or damage. The Seller, however, purchases the cargo insurance.

- **DAT**—Delivered At Terminal (... named terminal at port or place of destination). The Seller delivers when the goods, once unloaded from the arriving means of transport, are placed at the Buyer's disposal at a named terminal at the named port or place of destination. "Terminal" includes any place, whether covered or not, such as a quay; warehouse; container yard; or road, rail, or air cargo terminal. The Seller bears all risks involved in bringing the goods to and unloading them at the terminal at the named port or place of destination.

- **DAP**—Delivered At Place (... named place of destination). The Seller delivers when the goods are placed at the Buyer's disposal on the arriving means of transport ready for unloading at the named place of destination. The Seller bears all risks involved in bringing the goods to the named place.

- **DDP**—Delivered Duty Paid (... named place). The Seller delivers the goods—cleared for import—to the Buyer at the destination. The Seller bears all costs and risks of moving the goods to their destination, including the payment of Customs duties and taxes.

Terms Available Only for Maritime

In addition, several commonly used acronyms describe maritime/ocean transportation situations. These include the following:

- **FAS**—Free Alongside Ship (… named port of shipment). The Seller delivers the goods to the origin port. From that point, the Buyer bears all costs and risks of loss or damage.
- **FOB**—Free On Board (… named port of shipment). The Seller delivers the goods on board the ship and clears the goods for export. From that point, the Buyer bears all costs and risks of loss or damage.
- **CFR**—Cost and FReight (… named port of destination). The Seller clears the goods for export and pays the costs of moving the goods to destination. The Buyer bears all risks of loss or damage.
- **CIF**—Cost Insurance and Freight (… named port of destination). The Seller clears the goods for export and pays the costs of moving the goods to the port of destination. The Buyer bears all risks of loss or damage. The Seller, however, purchases the cargo insurance.

Selecting the proper INCOTERM for each global shipment can have a dramatic and positive impact on the capability to effectively balance the international transportation responsibilities between the exporter and the importer. Although there are a number of factors that affect INCOTERM selection, some of the most critical determinants typically include the relative expertise of each firm involved in the transaction, each firm's willingness to perform the required tasks, the type of product being sold, the mode of transportation being used, and the level of trust between the firms.

In addition to INCOTERMS, following is a sample of some of the additional documents that are common to complete a global business transaction. This is by no means an all-inclusive list, but rather a small subset of some of the most common and critical documents.

Export Documents

When supply chain managers export goods, three types of documentation are used to validate the shipments. These are the following:

- **Export licenses**—Government documents that authorize the export of strategic goods in specific quantities to a particular destination. This helps the government to maintain control over certain types of goods.

- **Shipper's export declaration**—Used to control exports and act as a source document for official export statistics. In the United States, this document is prepared by the exporter for all shipments exceeding a nominal value and is presented by the transportation service provider for presentation to the U.S. Customs and Border Protection at the port of export.
- **Certificate of end use**—Intended to assure authorities in the exporting country that the product will be used for legitimate purposes. End-use certificates are typically provided by the importing country's government.

Transportation Documents

Several types of documentation are used to track and verify ground shipments, including the following:

- **Freight manifests**—Internal carrier documents that list the exact makeup of the cargo, its ownership, port of origin and port of destination, handling instructions, and other key information. The accuracy of these documents is critical security regulation compliance.
- **Bills of lading**—Various types used for international shipments (such as ocean bill of lading, airway bill, through bill of lading, and intermodal bill of lading). Generally a bill of lading serves as a contract for carriage between the transportation company and the cargo owner who is either the exporter or importer depending on the INCOTERM used. As the contract, this document specifies the price and instructions for moving the freight. The bill of lading also serves as a receipt for the goods once received. When the transportation company signs the document, it acknowledges that it has received the cargo in good condition and in the correct quantity.
- **Packing list**—Provides a detailed inventory of the contents of a shipment. It itemizes the material in each individual package and indicates the type of package, such as a box, crate, drum, or carton. It also shows the individual net, legal, tare, and gross weights and measurements for each package.
- **Shipper's letter of instruction**—Spells out the requirements for handling in-transit goods. Additional documents may also be necessary, especially in the cases of shipping hazardous cargo.

Import Documents

Finally, when importing goods into a country, two other documents are necessary:

- **Certificate of origin**—A widely used and required import document. It is a statement that the goods were shipped from the country in which the exporter is located. The related certificate of manufacturer must also be signed by a commerce representative in the exporting manufacturer's country. These documents are often used by importing countries to help determine the necessary taxes, tariffs, and other regulations that may apply.

- **Certificate of inspection**—There are several varieties of these types of documents that are designed to attest to the authenticity, accuracy, quality, and safety of the goods. Each typically involves some type of an inspection process to ensure that goods conform to the description contained on the commercial invoice.

Route Planning

Route planning and delivery scheduling activities are not trivial, but instead often involve big dollars, affect customers, and can cause major headaches if not properly managed. Although they are not difficult problems to understand conceptually, they can be extremely challenging to solve. With the long distances and multiple route options involved in global transportation, the task of solving the route planning and scheduling dilemma is considerable.

Many supply chain managers might simply choose to leave the route planning and scheduling up to the carrier chosen to move the freight. Although this may be acceptable, the global transportation manager can hardly take a total hands-off approach and hope the function is performed adequately. In fact, they should be actively involved in the decision-making process because effective routing practices have the potential to affect customer satisfaction, supply chain performance, and organizational success. Transit time and on-time performance depend heavily on proper scheduling and sequencing of stops. Effective routing also helps avoid unfriendly countries, poorly equipped ports, and congested border crossing points that may drastically delay cargo flows.

Given the tremendous dollars spent on global transportation, efficiency is another major issue. The costs of global transportation services can easily top $1 trillion annually. With these costs at stake, carriers and their customers must develop highly efficient routes that maximize equipment capacity utilization. Effective collaboration between carriers and customers can help to optimize routes and schedules, minimize inefficiencies, and create a more cost-effective network that enhances customer service levels.

Under certain circumstances, one routing strategy that has gained popularity both domestically and globally is the use of intermodal transportation. For example, the capability to use multiple modes of transportation can open up alternative routes as opposed to being forced to use an all-water service. Intermodal transportation can take on several different forms and can help to create innovative solutions including a land bridge, mini-bridge or micro-bridge.

Land bridge transportation involves movement from one seaport to another using a combination of water and rail transportation modes. This routing method is effective in part because the goods remain in the same container and a single bill of lading covers the entire intermodal journey. A related routing option is the mini-bridge. This all-water alternative focuses on the combination of water and rail service with a port as the origin or destination point for the shipment. A micro-bridge is similar to a mini-bridge, with the main difference being the origin or destination point. This type of shipment originates or terminates at an inland port rather than an ocean port.

The most pressing issue in the intermodal transportation market is congestion. Although the ocean carriers are adding capacity to meet the growing demand levels, transfer points can quickly get clogged with freight. During peak demand periods, U.S. seaport facilities along the Pacific coast have struggled to keep product flowing through their facilities in a timely fashion. Intermodal capacity problems in the rail industry have also surfaced. Equipment shortages, transfer facility congestion, and labor issues create delivery delays and supply chain disruptions. As a result, these issues need to be evaluated as part of the intermodal routing decision-making process.

Product safety is also a concern when developing routes, particularly in certain global transportation situations. Some surface transportation areas are known to result in frequent hijackings, experience high levels of product theft, and even produce significant levels of product damage due to poor infrastructure. Although surface transportation has its challenges, other forms of transportation are not immune to trouble, either. Ocean carriers sail in areas known for piracy and where vessel attacks or seizures are not uncommon. Clearly paying close attention to cargo routing and scheduling can help to minimize these issues from negatively affecting the safety and security of products being transported around the globe.

Logistics Service Providers

Of course, transportation companies are not the only global logistics service providers (LSPs). LSPs also play a key role in the flow of goods between export and import locations. LSPs are external suppliers that perform all or part of a client's logistics functions. In the global logistics arena, activities can range from export packing to customs clearance and beyond. Some LSPs are specialists, handling just a few related logistics activities on a local or regional basis; others provide integrated services and one-stop shopping for all your global requirements. A summary of desirable capabilities for a modern 3PL are included in Figure 5-7.

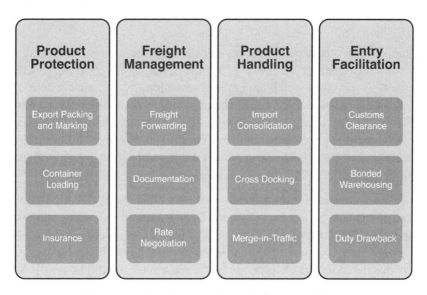

Figure 5-7 Logistics service provider capabilities

The growth of global sourcing and distribution has driven 3PL's pursuit of integrated capabilities and global reach. Customers' increasing activity in global sourcing and distribution has driven LSPs to bolster their international resources through the creation of internal divisions, acquisition of smaller LSPs, or the development of partner relationships with other LSPs. They are building logistics expertise and well-developed transportation networks to accommodate the growing volume of trade between key regions of the world.

Some of these organizations are investing in strategically located transportation and distribution facilities to address customers' global service requirements. These assets can help an organization establish critical hubs, streamline flows, and support customer fulfillment needs. In fact, some of these organizations have even established a physical presence in Asia, Europe, and Latin America, allowing them to serve a wide variety of countries and territories.

Strategies and processes for selecting the types of external organizations that can manage these types of supply chain facilitating arrangements are similar to those for making the carrier decision. It is necessary to carefully evaluate potential service providers and select those whose capabilities, commitment level, and price match the buyer's requirements. Comfort with the service provider's team is another key consideration to examine. This can be a time-consuming process, but it will greatly increase the likelihood of a mutually beneficial and successful relationship.

Summary

Effective supply chain management is a critically important component of almost any successful twentieth century business. The importance of effective supply chain management practices may very well be magnified when looking at today's successful global businesses. If performed properly, globalization offers companies the opportunity to simultaneously grow revenues and decrease costs (Chopra & Meindl, 2010). This can be an attractive opportunity for many companies to pursue. However, building a successful global business complete with the design and execution of a global supply chain network is quite challenging.

This chapter has addressed some of the key elements, challenges, and opportunities associated with a global business enterprise. Each business situation is somewhat unique and may require some customization to successfully achieve the desired outcomes. Nevertheless, several of the key common attributes faced by most global business operations have been briefly discussed in this chapter. Each of the main topics and issues discussed may require significant study and evaluation prior to any global supply chain manager being able to set up a network designed to achieve the desired level of success.

References

Bordner, T. (2013)The new world of global trade management. *Supply Chain Brain*. Retrieved September 3, 2013, from http://www.supplychainbrain.com/content/nc/technology-solutions/global-trade-management/single-article-page/article/the-new-world-of-global-trade-management/.

Chopra, S., and Meindl, P. (2013) *Supply chain management: Strategy, planning, and operation*. Upper Saddle River, NJ: Pearson Education.

Coyle, J. J., Langley, C. J., Novack, R. A., and Gibson, B. J. (2013) *Supply chain management: A logistics perspective*. Mason, OH: South-Western Cengage Learning.

Diederichs, R., Leopoldseder, M. (2008) It's still a big world. Retrieved August 1, 2013, from www.doi.org.

Dornier, P., Ernst, R., Fender, M., and Kouvelis, P. (1998) *Global operations and logistics: Text and cases* (1st ed) New York: John Wiley and Sons.

Hausman, W. H., Lee, H. L., Napier, G. R. F., and Thompson, A. (2009) *How enterprises and trading partners gain from global trade management*. Retrieved September 3, 2013, from http://www.scdigest.com/assets/reps/Stanford_GTM_Report.pdf.

Keegan, W. J. (2013) *Global marketing management* (7th ed). Upper Saddle River, NJ: Pearson Education.

Lee, H. L. (2004, October) The triple-a supply chain. *Harvard Business Review*, 1–11.

Novack, R. A., Gibson, B. J., and Bardi, E. J. (2011) *Transportation: A supply chain perspective* (7th ed). Mason, OH: South-Western Cengage Learning.

Simchi-Levi, D., Kaminsky, P., and Simchi-Levi, E. (2008) *Designing and managing the supply chain: Concepts, strategies, and case studies.* New York: McGraw-Hill/Irwin.

Wood, D. F., Barone, A., Murphy, P., and Wardlow, D. (2002) *International logistics* (2nd ed). New York: AMACOM.

6

WORLD CLASS SUPPLY CHAIN PERFORMANCE

World class supply chains cost-efficiently source, produce, and deliver thousands of products to customers every day while achieving high standards for reliability and customer service. Supply chains exist in complex, rapidly evolving environments. Much of the complexity comes from the fact that almost all supply chains consist of collections of multiple companies working together toward the achievement of mutually established goals (Mentzer, 2004). Determining how effective supply chain companies are in achieving these goals is the role of performance measurement.

This chapter explains the need for supply chain performance measurement and provide the rationale that companies use when creating their supply chain metrics. The chapter is organized into five sections. The first section explains the role of measurement in supply chains and distinguish between qualitative and quantitative measures. The second section describes the trade-offs to be considered when choosing between efficient, effective, and adaptable supply chain capabilities. The type of measures found in supply chains are discussed next, along with the characteristics of good measures. The final two sections describe three popular measurement frameworks and describe an approach to financial analysis facilitated by the strategic profit model.

Role of Measurement

Performance measurement is the evaluation of how well previously established goals have been met (Mentzer & Konrad, 1991). Performance measurement may be used to identify problem areas requiring immediate attention. Thus, measurement provides a means for coaching employees in corrective techniques and pinpointing process problems requiring a design improvement. Performance measurement also highlights outstanding results so that the participants involved can be recognized. Thus, measurement can be leveraged to reinforce the "right" actions and enhance motivation in an organization.

The evaluation of performance is often difficult because it must be seen across multiple dimensions (Chow, Heaver, & Henriksson, 1994) and includes the activities of multiple organizations that frequently have conflicting goals (Chow et al., 1994). Even closely aligned organizations in a supply chain may avoid sharing specific performance information, forcing managers to use internal, firm-specific performance measures as a surrogate for supply chain performance (Lambert & Pohlen, 2001).

In fact, the majority of supply chain metrics being used today are focused on the performance of a single company or the results of a dyadic collaboration (Fabbe-Costes & Jahre, 2008). The frequent use of internal metrics as a substitute for supply chain performance measures isn't surprising because supply chain managers are compensated on their ability to achieve their own company's goals and not the broader goals of the end-to-end supply chain.

Types of Performance Measures

Performance measures come in two broad varieties: quantitative and qualitative. Both types of measure are valuable, but one is much more applicable for use in assessing supply chain performance.

Qualitative measures are sometimes referred to as "soft" measures because they are used to classify results across several categories (Chow et al., 1994). Qualitative measures of performance (for example, good, adequate, and poor) provide very general feedback. Because they are not precise, qualitative measures provide limited guidance managers can use to fix existing problems (Beamon, 1999). These types of metrics are often used to capture customer perceptions. The data used to drive these measures is collected using a survey technique.

For example, customers might be asked, "How do you rate your most recent purchase experience with our company?" and then given a 5-point scale to respond to with 1 scored as poor, 3 scored as acceptable, and 5 scored as excellent. These qualitative responses can be quantified to some degree by assigning the 1–5 value to each response, summing these values across all responses, and then dividing by the number of responses, as shown in Figure 6-1. Although it is useful for supply chain managers to know that customers rate the service received as 4.1 on a 5-point scale, the steps necessary to improve the service in the future and increase that rating are not clear.

Quantitative measures involve the calculation of "hard" financial or operating numbers producing a numeric outcome in one of several forms, including a flat number (10 shipments were damaged in transit last week), a percentage (99.5 percent of orders were picked accurately), or a ratio (Asset Turnover is 2.2 this quarter). Each of the figures contains significant information about the degree of performance and quality of service provided. When compared against past performance results or established performance standards, quantitative metrics also directly highlight areas requiring immediate attention and improvement.

Using the following scale, rate your most recent purchase experience with our company.

Poor   Acceptable   Excellent
☐   ☐   ☐   ☐   ☐

Calculating the Metric:

	Poor		Acceptable		Excellent	Totals
Response Value	1	2	3	4	5	
Count	4	8	35	68	83	198
Product	4	16	105	272	415	812

Average   812 / 198 = <u>4.1</u>

Figure 6-1   Quantifying qualitative analysis

Supply chain processes lend themselves to evaluation using quantitative techniques. The result of most supply chain processes can be classified as a binary (good/bad) outcome. Here are a few examples:

- Manufacturing

  Number of units produced

  Percentage of units rejected as defective

- Distribution

  Number of picks per man-hour

  Number of returns

  Percentage of orders shipped same day they were received

- Transportation

  Number of miles driven this month

  Percentage of deliveries returned as damaged (in transit)

Another way to view supply chain performance is through a financial lens. What did it cost to store a pallet of product or ship a carton across the country? What return did the company earn on its investment in new material handling equipment? Ease of understanding and actionability are reasons why quantitative measures of performance are the preferred metric design for evaluating supply chain activities (Beamon, 1999).

Trade-Off Analysis

Supply chain executives are given the challenge of achieving and routinely improving upon an assortment of goals. The challenge is that many of these goals are, by their nature, competing; good performance in one area requires performance in another area to be reduced. Consider the following examples of the type of competing decisions supply chain managers routinely face:

- **Improve customer service**—Lower the cost per item shipped
- **Improve on-time delivery**—Reduce transportation expense
- **Reduce packaging expense**—Reduce the amount of damaged product customers receive

Many more examples are possible. The point is that supply chain managers are constantly confronted with the need to drive continuous improvement throughout their operations. These challenges are not unique to the supply chain; all areas of a business are pressured to show improvement. The difference from a supply chain perspective is that supply chain errors directly affect customers, the lifeblood of the company.

Thus, trade-offs are inherent in supply chain activities, but reducing performance in one area to accommodate results elsewhere is generally viewed as unacceptable. This section will look at the broad capabilities incorporated in most supply chains and discuss the importance of balancing the outcome of each.

Focusing the Supply Chain

Companies define strategy to guide their decisions and ensure that the organization has a clear approach for competing in the marketplace. Supply chain operations need to produce financial and customer service outcomes that align with company direction. Strategic alignment is assured when operations are fine-tuned to optimize performance in one or more of three broad areas. They include focusing supply chain processes to emphasize strong performance in the areas of efficiency, effectiveness, and adaptability highlighted in Figure 6-2.

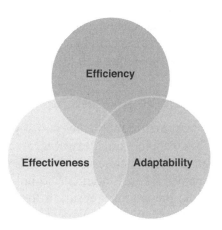

Figure 6-2   Performance trade-offs

Supply Chain Efficiency

Efficient performance is the measure of how well resources expended are utilized (Mentzer & Konrad, 1991) and is most often equated with the capability of the supply chain to provide the required level of service at the lowest cost (Mentzer, 2004). Efficiency-oriented supply chains strive to achieve low-cost operations. It has been argued that most supply chains focus on creating low cost operations, perhaps to their long-term detriment (Lee, 2004).

Optimizing supply chain operations from an efficiency perspective focuses on creating the capability of the supply chain to provide a required level of service at the lowest cost (Mentzer, 2004). In times of rising costs, supply chains emphasizing efficient outcomes may choose to reduce service in order to maintain cost. Whether it is the main point of emphasis, it should be assumed that all supply chain operations retain a level of interest in achieving efficient performance.

Supply Chain Effectiveness

Effective performance is a measure of the gap between customer expectations of performance and customer perceptions of the quality of the actual service delivered (Sharma, Grewal, & Levy 1995). Effectiveness-oriented supply chains strive to create high levels of customer service. A difficulty with measuring effectiveness is that it is problematic to observe because the true gauge of customer service falls outside the company's control; it comes from the mind of the customer in the form of customer satisfaction.

You should recognize that the qualitative measures discussed earlier are well-suited for determining supply chain effectiveness. Many supply chains also create internal measures of effectiveness as a proxy for external, customer-based measures that may be expensive and difficult to obtain. An example is the use of an on-time shipment metric. On-time shipment is easy to calculate for a company because the information needed is captured in-house:

On-time shipment = Number of orders given to carrier by cut-off / Total orders

Calculating an on-time shipment metric allows the company to know how it performed on the timeliness aspect of the customer's order experience—at least the portion of timeliness the company had control over. On-time shipment falls well short of being a viable substitute for customer satisfaction because it cannot account for transportation delays in the supply chain that occur after the shipment occurs, in-transit damage, the helpfulness of representatives on the customer support line, or any number of other things customers consider when they determine their degree of happiness with the service received.

Optimizing supply chain operations from an effectiveness perspective focuses on creating the ability of the supply chain to provide the maximum level of service for a given cost. Because the achievement of the highest levels of service can result in rapidly escalating costs, few companies can afford to focus solely on effectiveness outcomes.

Supply Chain Adaptability

The supply chain environment is subject to unanticipated and rapid changes. Customers desire new and improved service options, new competitors emerge, work stoppages slow the flow of raw materials, and a profusion of disruptions occur. (Wars, pirates, and weather have all had major impacts on supply chains in the recent years.) Companies that demonstrate the ability to rapidly adjust to the changing environment are set up for a long-term competitive advantage (Lee, 2004).

Optimizing supply chain operations from an adaptability perspective involves the use of skills outside of core supply chain capabilities that are the focus of efficiency or effectiveness excellence models. Adaptable firms possess outstanding market-sensing capabilities and identify shifts in the marketplace earlier than the typical firm. Adaptability often leverages the expertise of third parties and may require the creation of multiple supply chains (Lee, 2004). The adaptability-oriented company may not ever develop deep efficiency or effectiveness-related capabilities because this type of organization expects to tear down and remake its supply chain multiple times.

Understanding the Trade-Offs

All supply chains need to have goals—and thus metrics—that address the areas of efficiency, effectiveness, and adaptability. The needs of each situation are unique and there is no optimal trade-off point that can be pinpointed. Each company must determine which of the three performance elements need to be emphasized and which others should be deemphasized.

A company that emphasizes efficiency may discover that its supply chain has been so fine-tuned to the existing state of the world that it finds it extremely difficult to adapt to changing market conditions. In the extreme case, the company lacking an adaptability capability may see its costs rise rapidly as it struggles to accommodate change or demand for its products and services drop as customers flee to competitors more adept at meeting their needs. You can think through similar risks that exist for companies that overemphasize one of the other areas.

Most firms establish supply chain metrics that link to each of the strategic areas of efficiency, effectiveness, and adaptability. Depending on the strategic focus taken by a supply chain organization, the set of performance metrics used will need to be created to highlight the most valued characteristics of the operation. As you read through the next section of the chapter, think about which of the metrics described would work best for a firm focused on each of the three trade-off areas just discussed.

Types of Measures

The availability of powerful computer systems and the routine collection of large amounts of data have eased the burden of performance measurement for supply chain companies. Identifying the right measures still differentiates the best-run supply chain organizations from the rest of the pack, however. This section first provides an overview of the characteristics of good supply chain measures. Later, five broad categories of performance measurement are outlined, and multiple specific measures in each area are described.

Characteristics of Good Measures

Today, data is abundant and enables metrics to be easily created to gauge almost any aspect of performance. An issue that supply chain companies must guard against is the risk of creating more measures than can reasonably be monitored. Thus, fewer good measures are preferable to a large group of measures. How can good measures be determined? Table 6-1 provides a summary of 10 characteristics found in the best measures (Keebler, Manrodt, Durtsche, & Ledyard, 1999).

Table 6-1 Characteristics of Good Measures

A Good Measure	Description
Is quantitative	The measure can be expressed as an objective value.
Is easy to understand	The measure conveys at a glance what it is measuring and how it is derived.
Encourages appropriate behavior	The measure is balanced to reward productive behavior and discourage game playing.
Is visible	The effects of the measure are readily apparent to all involved in the process of being measured.
Is defined and mutually understood	The measure has been defined by and/or agreed to by all key process participants.
Encompasses both outputs and inputs	The measure integrates factors from all aspects of the process measured.
Measures only what is important	The measure focuses on a key performance indicator that is of real value to managing the process.
Is multidimensional	The measure is properly balanced between utilization, productivity, and performance and shows the trade-offs.
Uses economies of effort	The benefits of the measure outweigh the costs of collection.

Supply chain activities consist of a series of processes. The output or outcome of each process can be objectively measured using *quantitative* measures. Quantitative measures can be expressed as an objective value, making performance evaluation relatively easy. For example, a desired outcome of a picking process is the percentage of orders picked completely; obviously, other outcome measures are also of interest (such as the percentage of orders picked accurately and the percentage of orders picked and shipped on time). For purposes of this example, however, we will focus only on orders picked completely. If 952 out of 1,000 orders processed in a shift are judged to be picked completely, 95.2 percent of orders are deemed to have been completely fulfilled.

Easily understood measures consist of at least two components. First, the metric clearly conveys what is being measured and how it is calculated. In the example of the orders picked, the complete metric described previously, there is little ambiguity for the experienced reader what the result of 95.2 percent means. Second, measures are easily understood because they are defined and agreed upon by participants in the process being measured. Individuals tend to better understand a metric if they have been involved in its creation (Ferrell, 2007).

The measures utilized will influence the behavior of those participating in the process being evaluated. Appropriately defined metrics therefore *encourage appropriate behavior*. Alternatively, improperly defined or misaligned metrics will result in inappropriate behavior from key participants; appropriate and inappropriate in this context, meaning that the behavior either supports the achievement of supply chain goals or it doesn't. For example, a goal of 100 percent accuracy in order picking may be supported by a measure such as percentage of shipments audited. However, a measure focusing on percentage of orders shipped the same day may encourage participants to focus on picking fast rather than picking accurately.

Good measures are routinely available and *visible* to participants. Leading supply chain companies make metrics readily available to employees. Posting key metrics on a bulletin board next to the location where hourly employees clock in is one example of providing visible measurement. Posting metrics in an office area where employees rarely go or not posting metrics at all fails the visibility requirement and significantly diminishes their value of measurement.

Effective measures include both *outputs and inputs* of the process being measured, which allows both cause and effect to be measured and addressed. For example, a decreasing on-time delivery rate might be caused by late pickups, shipments not being ready on time, inaccurate delivery address information entered during order capture, or even production shutdowns. A 98 percent result on the on-time delivery metric supports many of the criteria already discussed (quantitative, easily understood, and so on), but it doesn't help diagnose the root cause(s) of the problem occurring with the 2 percent of shipments not delivered on time. A complementary metric that reports on the causes of late deliveries would be helpful in this case.

Companies that aren't sure what to measure may fall into the trap of attempting to measure too much. It is essential that supply chain companies *measure only what is important* to the success of each process. Measuring everything imaginable generally produces an unfocused workforce because it's impossible for participants to focus on more than a few key performance metrics. In some cases, the easy availability of data leads to the creation of metrics that do not help the company achieve primary goals. A frequent problem is the inability to eliminate old measures that have lost their usefulness. Often, metrics are created to address a specific problem. Unfortunately, once the problem is resolved, the metric remains and continues to be reported. Over the course of many months and years, this can result in the reporting of many meaningless metrics.

Supply chain performance is a complex concept that spans *multiple dimensions* including financial, operational, and customer service criteria (Chow et al., 1994). Excellent performance cannot be determined on a single dimension such as low cost because other dimensions, such as customer service, may be adversely affected by attempts to optimize a single dimension. A single metric cannot cover multiple dimensions, but the entire program of metrics produced must do so. Scorecarding (Brewer & Speh, 2000) is one technique (discussed later in this chapter) that has been used to capture multiple performance dimensions in one evaluation tool.

To be useful, supply chain metrics must be relatively economical to produce. Time and effort is required to collect data used to calculate metrics. If the effort to collect the data is greater than the benefit derived from learning the result, or if no positive active action is taken as a result of learning the metric outcome, the metric has no value and should be discontinued.

Good metrics should *facilitate trust* among the participants. Participants must trust that the metrics reported are accurate. This is especially true of metrics that cross organizational boundaries and involve the participation and support of multiple supply chain partners. Collaboration is a critical success factor in most supply chains, and the trust developed in performance measurement provides a solid foundation for collaborative activities.

Other Considerations for Supply Chain Measures

Avoid the average performance trap. Performance measures are highly repetitious: They are reported daily, weekly, and monthly. A technique found on many performance reports is to show average performance, or an expected standard performance goal. These comparison figures are useful, but can lull participants into a false sense of security. For example, a report may show that 95 percent of shipments leave the facility on time, which exceeds the existing standard. Often metrics with "acceptable" results are not evaluated as stringently as metrics highlighting poor results. Several questions should be asked. What's happening the other 5 percent of the time? Is this a cyclical problem? Are we negatively affecting customers 5 percent of the time? Minor problems lurking just under the surface may grow into major issues over time if they are allowed to continue unfettered.

Raise the performance bar over time. Customer expectations rise over time. Competitors improve their capabilities. Companies cannot afford to remain the same when last year's industry-leading performance may become average performance. Continuous improvement is a central tenet in today's best supply chains and must be reflected in the performance standards identified with key metrics.

Metrics must evolve over time. Processes change over time to accommodate changing customer needs and achieve the goals of continuous improvement initiatives. Unfortunately, many companies fail to consider changing their supply chain metrics to stay in step with these changing processes. Simply put, measures must evolve and adapt as goals and processes change to address shifting market priorities. Also, as stated earlier, measures with little current value need to be eliminated periodically.

The customer's point of view is the most important. This is one of the most frequently overlooked aspects of establishing supply chain metrics. Most companies focus on producing metrics they can control: orders picked and shipped complete and on time, in-stock availability, inventory turnover, and so on. Certainly these types of metrics are critically important to running an efficient and effective supply chain operation, but operations performance ultimately doesn't matter if the customer is unhappy with the result of the transaction.

A small but growing number of companies are establishing key metrics to determine the customers' perspective of supply chain performance. Although most supply chain organizations measure controllable outcomes such as percentage of orders shipped on time, the customer cares only about whether the order arrived on time, undamaged, and with proper documentation and accurate billing. Data for customer-driven metrics is more difficult to acquire, generally requiring some form of survey to be administered. The key question for supply chains is actually this: "Was the customer happy with the service provided by supply chain operations?"

Categorizing Supply Chain Performance Measures

Companies use a wide variety of supply chain metrics, and no universally accepted set of metrics exists. Thus, a comprehensive list of the supply chain metrics used is not provided, but a classification scheme is presented to describe the major types of measures used by many companies. In addition, multiple examples of metrics fitting each type are described in this section.

Figure 6-3 presents the supply chain metric classification scheme. The column categories contain metrics largely calculated from financial data. The five categories of cost, productivity, quality, asset management, and customer service have been described in previous research (Bowersox, Closs, & Cooper, 2002). These five categories are further delineated in Figure 6-3 with metrics from each being distinguished by whether they are customer-facing, supplier-facing, internal (company-specific), or holistic metrics (Hoffman, 2007). The classification scheme is described next.

Cost Performance

Cost is a reflection of the total funds spent on each functional area. Accounting systems traditionally accumulate costs at the department level that can be rolled up to the functional level to accommodate reporting of warehousing costs, transportation costs (inbound, outbound, and combined), order processing, inventory management (raw materials, finished goods, and so on), and other functional breakdowns that may be specific to the way a company is organized. It is becoming more typical to find costs reported at both the activity level (such as picking) and major process level (such as order fulfillment).

Perspective is provided by the reporting of cost data in relation to figures that reflect the size of the business such as cost as a percentage of sales or cost-per-unit of volume processed (for example, cartons, pallets, shipments, and so on). Reporting costs as a percentage of sales allows for easy benchmark comparison with published data available on an industry, or even a competitor in some cases (Hoffman & Barrett, 2010).

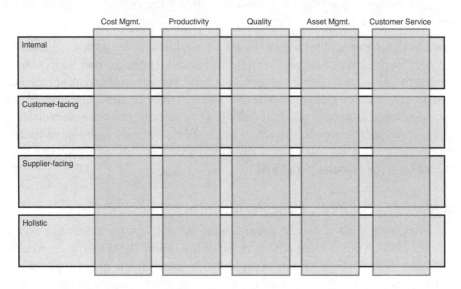

Figure 6-3 Metrics Classification Framework

Reporting costs in relation to key volume drivers provides perspective to participants and helps them understand the real cost of performing each activity across the supply chain. For example, transportation costs may be presented as cost-per-delivery, cost-per-mile, or cost-per-pound-shipped. Warehouse costs are often expressed relative to volume processed: cost-per-unit, cost-per-carton shipped, cost-per-pallet stored, and cost-per-shipment. This type of cost breakdown allows for easy comparison to historic performance (last month versus this month; last year versus this year) or performance standards. In either case, the information provided by this type of reporting helps identify areas where corrective action needs to be taken.

Cost reporting is not entirely focused on delineating costs to the lowest levels in the organization. Total landed cost is a key performance indicator for almost any supply chain. It is defined as the sum of all costs associated with a shipment, including purchase price, transportation charges, storage, insurance, and other related supply chain costs associated with procuring and moving a product from a point of origin to its final destination. An example of total landed cost from a company's perspective is shown in Table 6-2. All costs including purchase price from the supplier, all inbound transportation, material handling, storage, and inventory carrying costs are summed. This method of costing provides a more accurate picture of the true cost of making a product available for sale.

A possible list of cost-related metrics would be lengthy, to say the least, when you consider the number of activities in a given function and the different ways in which volume may be measured. An incomplete list of cost metrics includes the following:

- Total landed cost
- Cost as a percentage of sales
- Cost breakdown (including cost by function and cost by activity)
  Compare versus budget
  Compare versus standard
  Compare versus historical performance
  Weekly/monthly trend analysis
- Cost per unit/per carton/per shipment

Table 6-2 Total Landed Cost Example

Cost Factor	Amount (in $)
Purchase cost of one unit	100.00
Purchase order-related costs	1.00
Transportation (Ocean from Singapore-Port of LA)	0.25
Drayage	0.10
Transportation (Rail from LA-Memphis)	0.35
Warehousing (receiving, inventory mgmt., picking)	0.15
Inventory carrying costs (1 month)	0.05
Total landed cost	$101.90

Productivity

Productivity metrics express a relationship of inputs to outputs. Specifically, *productivity* measures how efficiently an output is created based on the quantity of inputs required to produce or accomplish it. Typical outputs in a supply chain differ based upon the echelon being considered. Table 6-3 provides several examples.

A potential difficulty in some cases is the inability to match up specific inputs to the outputs. An example is a company that captures outputs (e.g., orders shipped) daily but reports employee hours worked weekly. It will be easy for the example company to produce a weekly "shipments-per-employee-hour" metric because the daily number of shipments can be simply added up for any given week and divided by the total hours worked. The metric would be difficult, or impossible, to produce on a monthly basis because the input component isn't captured on a time scale that directly equates to a month (at least without requiring a great deal of manual effort to break weekly hours worked into days). Modern systems make this issue less of a concern than it was a few years ago, but it still remains a problem for many companies because the data used to produce many productivity metrics is captured across multiple systems. In this example, the order shipped may come from the order processing or warehouse management system, whereas employee hours worked typically comes from a payroll system.

Table 6-3 Typical Supply Chain Outputs to Be Measured

Supply Chain Echelon	Output
Raw materials supplier	Tons of raw material produced
	Number of customer shipments
Manufacturer	Number of units produced
	Number of customer shipments
Distributor	Number of orders filled
	Number of cartons shipped
Retailer	Number of customer sales transactions
Third-party carrier	Number of orders delivered
	Number of miles driven

Labor productivity is the primary focus on productivity metrics in most supply chain companies. Labor productivity may be evaluated in a variety of ways, including taking a macro view (total payroll cost or hours worked) or micro view (productivity of a group of employees or an individual employee). Equipment productivity is another frequently seen category of productivity metrics. Example productivity metrics include the following:

- Units/orders shipped (or manufactured, procured, delivered, picked, and so on)

  Per employee

  Per employee hour

  Per labor dollar

- Equipment utilization/equipment downtime

Quality Performance

Performance relative to product and/or service reliability reflects the degree of quality produced by a company's supply chain operations. Service reliability is typically measured for key outputs of each activity or function. Accuracy of supply chain activities is a routine quality metric and is calculated as the number of times an activity is performed correctly divided by the number of times the activity is performed. Examples of accuracy-driven quality metrics in a warehousing context include

- **Receiving accuracy**—Number of units received and correct stock-keeping unit (SKU) identified
- **Putaway accuracy**—Percentage of SKUs placed in correct bin location
- **Picking accuracy**—Lines picked correctly; quantity of units picked correctly

The accuracy of information stored in computer systems is another area of quality that must be monitored because inaccurate data actually creates inaccuracies in supply chain operations. Think about the impact on a warehouse operation of recording the wrong location for a SKU. This leads to incorrectly filled or unfilled orders and inaccurate stock-keeping. Or imagine a simple customer address error at the order entry, leading to unnecessary transportation costs to bring the shipment back to the Distribution Center (DC). Poor quality of data results in customer disappointment and additional cost to process returns.

Certain quality metrics allow the organization to determine its overall level of quality performance. Damage measurement is an example. Damage frequency is calculated as the number of damaged units divided by the total number of units in inventory currently, or the total number of units passing through the facility in a given time period. Damage frequency may be measured at each step of the supply chain process: unloading/receiving damage, storage damage, processing (picking and shipping) damage, and transportation damage. Often damage is not identified, however, until the customer receives an order and determines the item is damaged. This makes the returns process an important measuring stick for determining supply chain process quality.

Example quality metrics include these:

- Accuracy

    Order entry accuracy

    Receiving accuracy

    Stock-keeping (inventory location and quantity) accuracy

    Picking accuracy

    Information accuracy

    Billing accuracy

- Damage frequency

    Measured by specific activity

    Measured through returns

Asset Management Performance

Supply chain operations require significant capital investment. Capital investment into major assets with an anticipated long life, such as tractor trailers, buildings, and conveyor systems, are considered *fixed assets*. Short-term investments required to keep the business running are classified as *working capital*. The short-term nature of working capital is not to imply that these investments are minor. A great deal of a company's working capital is tied up in inventory; for wholesalers, inventory may represent 80 percent of total capital (Bowersox et al., 2002). Asset management metrics shed light on how well capital is utilized in supply chain operations.

A major capital investment for many companies is the facility network. Manufacturing plants, distribution centers, returns processing centers, and the like can be measured in terms of capacity utilization. Each facility will have an output, or throughput, expectation; and a capacity-utilization metric shows how much of that capacity is being used. For example, a warehouse capable of shipping 25,000 cartons per day but shipping only 15,000 cartons has a capacity utilization of 60 percent.

Capacity utilization may also be reflected in terms of time. A machine may have a capacity assigned to it based on the assumption that it will operate no more than 20 hours per day, with the other 4 hours set aside for machine maintenance. Thus, a capacity utilization in excess of 100 percent is possible and serves as a red flag to managers that the machine is at risk of breaking down due to overuse and lack of time allocated for proper maintenance. In this example, a result significantly less than 100 percent highlights unanticipated downtime.

Inventory turnover is the most frequently used indicator of inventory efficiency. This metric shows how frequently inventory is sold through and replaced. Inventory turnover is calculated as follows:

Inventory turnover = Cost of goods sold / Average inventory value at cost

This calculation reflects turns for a specified time period, such as month, quarter, or year. You may run across companies that use sales revenue in the numerator. Also, it is not unusual for companies to use the value of inventory at the end of the period being measured for the denominator. Slight differences in the calculation of turns tend to be industry-specific. When comparing turns between two (or more) companies, take care to ensure that the turns metric is calculated using the same formula for both companies.

An alternative inventory metric considers the amount of inventory on-hand relative to forecasted sales volume. An example of this inventory days of supply or days of sales metric is a SKU selling 10 units per day with 400 units in stock. This represents 40 days of supply for the SKU. A days-of-supply metric is useful in conjunction with lead-time analysis of replenishment purchases. The metric may not be useful for items with great variability in sales.

Return on assets (ROA) measures the profitability generated by the company's operational assets. ROA is calculated as follows:

Return on assets = Net profit margin / Asset turnover

Asset turnover is simply sales (from the income statement) divided by total assets (from the balance sheet). Asset turnover is a measure of how efficiently the company is utilizing its assets. The calculations for each component of ROA can be found in the strategic profit model discussion provided later in this chapter.

Example asset management performance metrics include these:

- Fixed assets

  Return on assets

  Asset turnover

  Capacity utilization

- Working capital

  Inventory turnover

  Inventory days of supply

Customer Service Performance

Customer service has been described as including the characteristics of availability, operational performance, and service reliability. Product availability is the capacity to have inventory in stock when desired by the customer. Availability can be reflected through the calculation of *fill rate*, which is often calculated at different levels, including the following:

Item fill rate = Number of items shipped / Total number of items ordered

Line fill rate = Number of order lines shipped complete / Total number of order lines

Order fill rate = Number of orders shipped complete / Total number of customer orders

Order fill rate is clearly the most stringent measure and will score lower than line fill rate or item fill rate, unless every order is completely filled, in which case all three fill rates shown here will be 100 percent. Companies routinely report the number of back orders and number of stockouts as complementary measures of availability that clarify the magnitude of an availability issue.

Order cycle time is a good measure of operational performance. The *order cycle* may be defined differently by different companies but is generally viewed as the elapsed time between the receipt of a customer order and delivery of the order to the customer. This may be reported as an average length of time. Order time variability is an important complementary metric that describes possible inconsistency in performance from the average. An average order cycle time of 6 days with a variability of +/− 1 day may be preferred by customers over a cycle time of 5 days with a variability of +/− 3 days because less variability suggests that the service provided is more reliable and allows the customer to plan more effectively.

Example customer service performance metrics include these:

- Fill rate
- Stockouts
- On-time delivery
- Order cycle time
- Order cycle time variability

Scope of Supply Chain Measures

Supply chain measures can additionally be classified by the scope of the individual metrics. This is represented graphically by the horizontal bars in metric classification framework (refer to Figure 6-3).

The vast majority of supply chain metrics being used today are focused on the performance of a single focal firm (Fabbe-Costes & Jahre, 2008). Although these *internal* metrics allow companies to manage their supply chain processes and resources effectively, they ignore the fact that supply chains involve the interactions of multiple firms to be successful (Mentzer, 2004).

True supply chain performance measurement requires the collection of performance data from across the many organizations participating in the supply chain. This is challenging in many cases because firms may be unwilling to share performance data because they view such data as strategic and proprietary. They fear that sharing performance-related data—even with collaborating partners—may expose their own weaknesses and vulnerabilities.

Dyadic metrics between two partnering firms, whether focused upstream (*supplier-facing*) or downstream (*customer-facing*), can be beneficial to the relationship of partner companies and improve aspects of supply chain performance for each. Fabbe-Costes and

Jahre (2008) found that about 15 percent of research studies they reviewed described the use of dyadic or intercompany collaborative metrics to measure specific areas such as cooperation, relationship commitment, service quality, and inventory visibility.

Holistic supply chain performance measurement is necessary for supply chains to achieve optimal results (Holmberg, 2000). Despite this admonition, supply chain-wide metrics that attempt to evaluate performance of entire supply chains from raw materials supply to the end consumer are exceedingly rare (Defee, 2012). Although holistic metrics are missing in most supply chains, a few key metrics used by retailers have been used as a proxy (Gibson, Defee, and Randall 2010). Retailer metrics are useful in this regard because they are the entity closest to the ultimate consumer where the culmination of overall supply chain efforts will be judged.

Measurement Systems and Frameworks

Supply chain performance is complex, dynamic, and multifaceted. The intertwined nature of multiple business processes and interaction of multiple organizations makes identifying and capturing appropriate metrics a challenging task.

Performance measurement is problematic because of the difficulty of getting quantifiable performance results from multiple member organizations (Brewer & Speh, 2000). Even when data is readily available, supply chain partners may not share common goals or pursue common outcomes (Chow et al., 1994). Although internal measures of operational performance have been used frequently (Lambert & Pohlen, 2001), true supply chain performance cannot be gauged unless a more holistic set of measures is used (Holmberg, 2000).

This section describes three methods of performance measurement that have proven useful when applied to the supply chain context. The three perspectives are not competing viewpoints but should be considered as lenses through which supply chain performance may be viewed.

Balanced Scorecard

The balanced scorecard was originally introduced as a tool for measuring company-wide performance (Kaplan & Norton, 1992). Brewer and Speh (2000) adapted the balanced scorecard approach and applied it to the supply chain environment almost a decade later. The balanced scorecard still stands out today as a solid tool for evaluating supply chain performance across multiple dimensions. As the name implies, performance must be "balanced." That is, the company does not want to excel in certain areas to the detriment of other areas.

The balanced scorecard uses performance measures across four dimensions: customer perspective, SCM goals perspective, innovation and learning perspective, and financial perspective (see Figure 6-4). The basic premise of the balanced scorecard is that the "framework balances the inclination to overemphasize [short-term] financial performance by incorporating metrics related to the underlying drivers of long-term profitability, namely, the business process measures, innovation and learning measures, and customer satisfaction measures" (Brewer & Speh, 2000, 83).

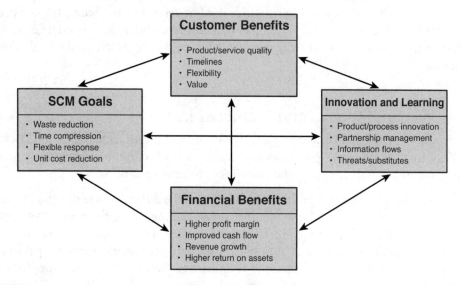

Figure 6-4   Supply Chain Balanced scorecard

A brief explanation of each of the four perspectives is provided here:

- **Customer benefits perspective**—The focus here is to determine what the customer must believe about the fulfillment system in order for the firm to be successful. The organization wants to measure this through customer value and retention. Customer value is measured through product and service quality, response time, and/or flexibility.

- **SCM goals perspective**—The focus is assessing the extent to which fulfillment processes meet customer needs. These measures are generally nonfinancial in nature. They may evaluate such dimensions as timeliness (order cycle time), flexibility (ability to rapidly meet customer order changes), ongoing cost reductions, and elimination of waste.

- **Innovation and learning perspective**—The question to be answered here is this: What do we need to do on a continuing basis to delight and retain customers? The focus is on future rather than current capabilities. Measures concentrate on factors such as the time required to place new products in the distribution pipeline, process improvement rates, and the extent to which data sets are shared across the supply chain.

- **Financial benefits perspective**—Traditional measures of financial success (for example, ROA metrics) must be met for the company's supply chain to remain viable. Profitability is evaluated by type of customer based on the cost to fulfill the requirements. The cash-to-cash cycle measures how long it takes the firm to convert funds spent on materials and labor into cash in-hand. The faster product moves through the supply chain, the faster cash will be generated for each supply chain member.

Sixteen specific measures are identified: four in each of the four performance dimensions. A difficulty with the balanced scorecard as originally proposed is that several of the measures require comparison against the performance of competing supply chains. Obtaining competitive data in several of the suggested areas may prove to be extremely difficult. The measures are listed as bullet points in Figure 6-4.

Another shortcoming is that a balanced scorecard approach does not provide a summary score for performance across all four areas. Although a net score produced from a balanced scorecard analysis might be useful in some cases, the value of the framework is that it pulls together key performance indicators across multiple dimensions into one place for easy and routine viewing. It is also doubtful that a net score would be meaningful when compared across multiple industries.

The bottom line of the balanced scorecard is that it provides a simple to understand framework that pulls multiple dimensions of performance into a single output. It should be viewed as a framework that can be adjusted to meet the needs of the organization and can evolve over time as those needs change in response to a changing marketplace.

Activity-Based Costing

Traditional accounting practice allocates costs to departments or cost centers in a business. These indirect costs may be allocated based on rules that simplify accounting for business costs such as allocating costs on the basis of labor hours used or machine hours recorded. But in reality, these types of allocations may not closely match the actual requirements of producing each output.

For example, products A and B may require the same number of machine hours to be produced, but product A may require additional manual customization on the back end of the process. If overhead costs are allocated strictly based on machine hours used, production costs will be understated for product A and overstated for product B.

In the process-oriented supply chain environment, the traditional approach is often not very useful because cost centers rarely align with the activities undertaken to accomplish supply chain tasks. Activity-based costing has proven to be a useful method for cost allocation in this respect.

Activity-based costing is an approach to understanding and controlling costs. In a nutshell, this systematic tool helps organizations build a realistic cost model for their operations. Activity-based costing is a technique that measures the cost and performance of activities, resources, and cost objects. The basic premise is that the demand for a company's outputs gives rise to the need for activities that in turn necessitate that resources be used (that is, costs incurred) in completing those activities. Resources (people's time, machine time, materials, and so on) are assigned to activities; then activities are assigned to cost objects based upon their use (Raffish & Turney, 1991). Indirect costs (that is, overhead allocations) are converted to direct costs assigned to specific activities.

Activity-based costing involved two primary procedures that are highlighted in Figure 6-5. First, resources used must be traced to the activities that require consumption of the resources. This will help determine the cost of performing an activity. Second, identify the outputs that require activities to be completed. This allows the activity costs to be linked to the products, services, or customers that made the activity necessary.

The goal of activity-based costing is to identify outputs (for example, products and services) that are not profitable. The knowledge gained from an activity-based costing analysis allows the firm to change an existing process to become more efficient and/or increase pricing to allow the product or service to become profitable. In the extreme case, the process/product may be eliminated.

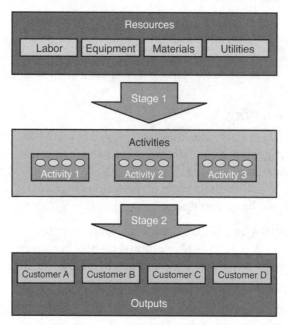

Figure 6-5   Activity-based costing

A downside of the activity-based costing approach is the data capture requirement. Systems must be in place to record activities as they occur, and even the cost of downtime when operations are not underway (Drucker, 1999). However, activity-based costing provides organizations with accurate and relevant cost information necessary to support sound business decisions. The results of activity-based costing allow the company to evaluate the financial impact of specific supply chain processes, identify opportunities to improve efficiencies, and develop new processes that are cost-effective. Activity-based costing is being adopted into many aspects of supply chain operations in an effort to more precisely understand what the true costs are of serving selected customers or performing certain activities.

Six Sigma DMAIC

Motorola Corporation developed the Sig Sigma quality improvement approach in the early 1980s (Tennant, 2001). The method became popularized following its adoption by General Electric in the 1990s. This continuous improvement approach utilizes statistical methods to focus on the identification and subsequent elimination of the causes of process defects (Pande, Neuman, & Cavanaugh, 2001).

The improvement of existing business processes are addressed using the DMAIC project methodology. A DMAIC project consists of five steps that are always followed in order. Once the problem is identified and understood (Define), data is collected (Measure) and analyzed to identify root causes (Analyze). Process improvements are developed and implemented (Improve), which then must be routinely measured to ensure that quality results are maintained over time (Control). The five phases of a DMAIC project are as follows:

- **Define**—Clearly describe the business problem; specify the customer set; and identify project goals, resources, and timeline.
- **Measure**—Identify key elements of the current process, determine the best way to measure each element, and collect data.
- **Analyze**—Ascertain gaps in actual versus expected performance. Conduct root cause analysis on each defect under investigation to determine the reason for each performance shortfall.
- **Improve**—Optimize the current process based upon data analysis. Design potential solution options, test solutions, and implement solutions providing the greatest improvement.
- **Control**—Produce a monitoring plan to evaluate the revised, future state process to ensure that any deviations from target are corrected before they result in defects. Maintain records of ongoing process evaluation (and additional changes) and ensure the workforce is appropriately and routinely trained.

DMAIC should not be confused with DMADV (Define, Measure, Analyze, Design, and Verify) projects intended for the creation of entirely new processes designed to achieve Six Sigma quality results.

Financial Outcomes

The value provided by supply chain management (SCM) is ultimately gauged through the assessment of financial performance. The most successful supply chains drive greater profitability for companies that play both central and supporting roles in supply chain operations. Profit improvement is found through a combination of enhanced customer service (effectiveness) and low cost of operations (efficiency).

Three critical elements are necessary to ensure that supply chain performance is measured appropriately. First, performance must be tied to supply chain strategy and goals. The existence of shared, measurable goals reduces the opportunity for managers to optimize their own organization's performance at the expense of the broader supply chain (Lambert & Pohlen 2001).

Second, and closely tied to point one, is the importance of developing holistic supply chain goals and metrics. Measurement limited to a level less than the entire supply chain will result in suboptimized performance as solutions are created to address problems identified through nonholistic measures (Defee, Stank, & Esper, 2010). Third, successful performance evaluation is multidimensional and requires coverage of at least efficiency and effectiveness. The strategic profit model provides a method that addresses each of these requirements.

Strategic Profit Model

Companies evaluate their performance through their capability to produce financial results and increase shareholder value (Beamon, 1999). The strategic profit model, presented graphically in Chapter 1, provides a method of analysis combining information from the income statement and balance sheet to develop key performance measures. Efficiency is measured through ROA. Effectiveness is measured through increased sales volume. The arrows corresponding to each financial element in the model identify the direction of change in the financial numbers needed to improve efficiency. Efficiency and effectiveness metrics developed from the strategic profit model are explained next.

Measuring Supply Chain Efficiency

ROA is the efficiency metric produced by the strategic profit model. ROA is enhanced by increasing net income faster than total assets. This is accomplished by one or more of several options, including increasing sales; reducing cost of goods sold (COGS); reducing sales, general and administrative (SG&A) expenses; or reducing assets of any type.

Operating costs, including COGS and SG&A, may actually rise, but as long as sales increase at a faster rate, gross profit margin and net income will increase, and even a minor increase in net income will produce an improvement in ROA as long as total assets remain constant.

When sales are not increasing, the only way to improve ROA is by creating greater operating efficiency or reducing the asset base. For example, operating efficiency is improved by purchasing raw materials at a reduced cost, or reducing the cost of selling and distributing (such as reducing the size of the sales force or shipping product using a less expensive method). Reducing total assets has a similar effect. This may be accomplished by lowering inventory levels, collecting outstanding receivables more rapidly, or selling long-term assets used in the production process.

Measuring Supply Chain Effectiveness

Sales (or more specifically increases in sales) is the effectiveness metric found on the strategic profit model. Effectiveness in the model is driven by delivering improved customer service, and growing sales are a direct result of increased customer satisfaction. Sales growth may occur because existing customers choose to shift purchases from competitors or because new customers are acquired. Sales growth improves ROA, so increasing effectiveness may also have a direct, positive impact on supply chain efficiency.

The approach to measuring efficiency and effectiveness described previously readily accommodates performance measurement for a single organization. These metrics can also be summed across all organizations in a supply chain to determine holistic supply chain performance. Of course, this assumes that the various companies that make up the supply chain have previously developed strong, trustworthy relationships and are prepared to share sensitive financial data.

Chapter Summary

As has been clearly stated, the supply chain environment is a complex one. Making sense of performance out of that complexity is the role of performance measurement. Selecting the type of metrics to use in evaluating the performance of a supply chain is a critical step toward success for supply chain managers, organizations, and the end-to-end supply chain enterprise.

The purpose of this chapter was to outline the requirements necessary for the creation of effective supply chain performance metrics. Effective supply chain metrics should be linked to company goals and the end-to-end supply chain enterprise. Supply chain managers are constantly challenged to balance among competing priorities such as low cost versus high customer service versus adaptability of service delivery. These trade-offs highlight the point that supply chain metrics must cover multiple dimensions of performance. These dimensions include performance reporting in the areas of cost management, productivity, quality, asset management, and customer service. Because supply chains involve the seamless interactions of many organizations, a complete program of supply chain metrics should include the performance of multiple firms. Finally, supply chain metrics should adapt as the supply chain continues to evolve. New metrics need to be created as new goals are established and new services are provided. Likewise, existing metrics need to be retired when their use has waned.

World-class supply chain organizations routinely fail at one or more aspects of their performance assessment programs. These organizations use these failures as an opportunity to identify and address process, communication, and integration problems. The key to successful performance assessment is ultimately the organization's capability to rapidly resolve issues, adjust, and adapt as problems occur.

References

Beamon, B. M. (1999) Measuring supply chain performance. *International Journal of Operations and Production Management*, 19(3), 275–292.

Bowersox, D. J., Closs, D. J., and Cooper, M. B. (2002) *Supply chain logistics management*. New York: McGraw-Hill.

Brewer, P. C. and Speh, T. W. (2000) Using the balanced scorecard to measure supply chain performance. *Journal of Business Logistics*, 21(1), 75–95.

Chow, G., Heaver, T. D., and Henrikkson, L. E. (1994) Logistics performance: Definition and measurement. *International Journal of Physical Distribution & Logistics Management*, 24(1), 17–28.

Defee, C. C. (2012) Do companies care about supply chain performance? Working paper.

Defee, C. C., Stank, T. P., and Esper, T. L. (2010) The performance implications of transformational supply chain leadership and followership. *The International Journal of Physical Distribution and Logistics Management*, 40(10), 763–791.

Drucker P. F. (1999) *Management challenges of the 21st century*. New York: Harper Business.

Fabbe-Costes, N. and Jahre, M. (2008) Supply chain integration and performance: A review of the evidence. *International Journal of Logistics Management*, 19(2), 130–154.

Ferrell, W. G. (2007) Time to rethink performance measurement. *Pensions & Investments*, 5(18), 31.

Gibson, B. J., Defee, C. C., and Randall, W. S. (2010) *The state of the retail supply chain: Results and findings of the 2009 study*. Retail Industry Leaders Association: Arlington, VA.

Hoffman, D. (2007) Supply chain measurement: Turning data into action. *Supply Chain Management Review*, 11(6), 20–26.

Hoffman, D. and Barrett, J. (2010) Benchmarking your supply chain. *The Journal of Commerce*, 11(35), 30–33.

Holmberg, S. (2000) A systems perspective on supply chain measurements. *International Journal of Physical Distribution and Logistics Management*, 30(10), 847–68.

Kaplan, R. S. and Norton. D. P. (1992) The balanced scorecard: Measures that drive performance. *Harvard Business Review*, 70(1), 71–79.

Keebler, J. S., Manrodt, K. B., Durtsche, D. A., and Ledyard, D. (1999) *Keeping score: Measuring the business value of logistics in supply chains*. Oak Brook, IL: Council of Logistics Management.

Lambert, D. M. and Pohlen, T. L. (2001) Supply chain metrics. *The International Journal of Logistics Management*, 12(1), 1–19.

Lee, H. L. (2004) The Triple-A supply chain. *Harvard Business Review*, 82(10), 102–112.

Mentzer, J. T. (2004) *Fundamentals of supply chain management: Twelve drivers of competitive advantage*. Thousand Oaks, CA: Sage.

Mentzer, J. T. and Konrad, B. P. (1991) An efficiency/effectiveness approach to logistics performance analysis. *Journal of Business Logistics*, 12(1), 33–63.

Pande, P., Neuman, R. and Cavanagh, R. (2001) *The Six Sigma way*. New York: McGraw-Hill.

Raffish, N. and Turney, P. (1991) Glossary of activity-based management. *Journal of Cost Management*, 5(3), 53–64.

Sharma, A., Grewal, D., and Levy, M. (1995) The customer satisfaction/logistics interface. *Journal of Business Logistics*, 16(2), 1–22.

Tennant, G. (2001) *Six Sigma: SPC and TQM in manufacturing and services*. Hampshire, England: Gower Publishing.

7

THE SUPPLY CHAIN OF THE FUTURE

The supply chain principles, processes, and strategies discussed throughout this book focus on supply chain structures, strategies, and processes that are being used today by leading-edge organizations. They realize that supply chain management (SCM) offers the opportunity to help drive profits through cost reduction and revenue contribution. As a result, SCM is helping them achieve shareholder value and build a competitive advantage in the marketplace.

Intelligent organizations are also attuned to the importance of being vigilant and forward thinking. SCM takes place in a vibrant, global environment in which strategies can lose relevance, networks can be disrupted, and cost dynamics can shift. Savvy managers continually analyze the events, technologies, and strategies that will shape supply chain strategy and practice. The ability to quickly adapt to change will further separate these leaders from the pack.

The concluding sections of this supply chain concepts book review ten major issues that will drive change and create challenges in supply chains in the coming years. Though the list is by no means comprehensive, each issue warrants the attention of supply chain managers. These driving forces include a variety of people, process, and technology considerations that will influence the success of the supply chain, the organization, and its stakeholders.

Ten Driving Forces

A number of transformative forces impact supply chains today, and these trends are expected to last well into the future. Managers should be aware of the potentially disruptive impacts of the following ten forces acting on the supply chain.

Driver 1: Permanent Volatility

With the memories of a recent global economic meltdown and epic natural disasters fresh in mind, supply chain managers realize that major problems can surface swiftly and cause great disruptions to product flows. Recent research by PwC (2013) notes that the macroeconomic cycles of growth, contraction, and recovery have become erratic, making reliable end-to-end supply and demand planning increasingly challenging. Another study points to disruptive macroeconomic factors of population growth and migration, rising economies, global connectivity, geopolitical activity, and environmental change that will lead to great frustration for supply chains and the managers who must adapt to the challenges (Autry, Goldsby, & Bell, 2012).

Accenture incorporates many of these challenges, along with shorter commodity cycles, supplier variability, and demanding customers, into a collective conundrum that it calls permanent volatility. Volatility and variability are not new issues to supply chain managers. What has changed is the increased velocity of challenges and frequency of volatility over the past 5 years. When coupled with the growth of globalization, a volatile environment emerges that requires a response from supply chain managers (Rasmus, Godfrey, & Richter, 2012).

One response is to use the advanced connective technologies discussed in Chapter 4. The ability to generate accurate, real-time data will help an organization assimilate business data into rapid and decisive responses. The result is better insight to action where the organization can predict and respond to changes in a timely fashion rather than react slowly (Rasmus et al., 2012). Despite their availability and growing interest, many companies are not yet taking advantage of the existing suite of innovative supply chain technologies such as radio-frequency identification (RFID), visibility tools, and statistical decision technology to facilitate process transparency, automation, and efficiency (PwC, 2013).

Another proactive strategy to temper volatility is the development of adaptable structures. To capitalize on new opportunities, traditional operating models and network structures that optimize labor costs must give way to a total landed cost view that considers inventory, transportation, lead times, and opportunity costs (Cudahy, George, Godfrey, & Rollman, 2012). Virtual supply chains and multilocation networks with the capability to rapidly align inventory with demand changes at low cost provide needed adaptability.

Agile execution of sourcing, production, and fulfillment operations allows a supply chain to manage volatility. Supply chains must be able to rapidly adjust to changes in supply and demand. A truly agile operation relies on its Human Resources being proactive and able to innovate, and its IT infrastructures and systems helping to detect and respond to opportunities. Being able to take an idea to market more quickly than a competitor, and to take full advantage of new technologies to get closer to key customers, is vital to future growth in today's marketplace (Ernst & Young, 2011).

Supply chain innovation is another critical capability for managing volatility in terms of shrinking product life cycles and changing demand opportunities. Organizations must design product and service bundles that meet the current and future needs of profitable customer segments. They must also recognize unprofitable products and customer segments and have the discipline to walk away from this business (Autry et al., 2012).

Volatility is a constant presence in the twenty-first century. Tremendous risks exist for organizations that fail to recognize the potential impact of permanent volatility and take steps to alleviate it. In contrast, organizations that develop adaptable, agile, and innovative supply chain operating models will limit their risk and capitalize on new opportunities.

Driver 2: C-Level Engagement

As a discipline, SCM is in the midst of a high-profile transition. C-level executives increasingly view the discipline as a strategic asset (Burnson, 2013b). They recognize the value of strong, integrated supply chain capabilities for propelling their businesses to greater levels of success, particularly in light of permanent volatility. These executives are elevating supply chain leaders to strategic roles and are investing in the supply chain to create competitive advantage.

From manufacturers to retailers and restaurants, savvy C-level executives recognize the capability of SCM to work across functions, control costs, ensure product availability, and affect the financial statement in good times and difficult times. In a recent conference panel discussion, the chairman and CEO of casual dining company Brinker International noted, "The fact is, supply chain is responsible for sourcing everything from raw materials to doing contract provisions. They handle everything you need for your restaurants and make sure you never run out or run long. Your supply chain is the key to a successful operation" (National Restaurant Association, 2013).

The good news is that this increased focus on SCM creates valuable opportunities for engagement. When included in executive level discussions, supply chain leaders can increase understanding of supply chain capabilities, shape strategy, and contribute to organizational success. The key is to communicate effectively with C-level leaders. That means clarifying supply chain opportunities, needs, and risks while avoiding supply chain jargon. It also means using readily understandable metrics to highlight performance. Most importantly, it means becoming conversant in the language of money and being able to articulate the value of SCM to C-level executives through financial statements and reports (Lee, 2013).

While the increased attention is generally positive for supply chain managers, it also means that the spotlight will burn brightly on missed opportunities. For example, senior executives believe that there is much room for supply chain improvement; particularly the area of supply chain risk management is a primary concern. A Deloitte survey of global executives revealed that senior executives are concerned about the growing complexity of global supply chains and the resulting frequency and expense of disruptions. Nearly one-half of the respondents indicated that their supply chain risk management programs are only somewhat effective or not effective at all. To address these concerns and improve risk management, supply chain managers must adopt a holistic and integrated process that includes tools such as predictive modeling. This will drive more proactive management of risk and the level of resilience that C-level executives desire (Burnson, 2013a).

This emerging opportunity must not be missed by supply chain managers. They must be prepared to sell the story of SCM to the increasingly receptive C-level executive audience. Of course, it will be essential to performance as promised because supply chain managers will be measured against the C-level executives' increasing expectations.

Driver 3: Talent Management

As C-level executives attempt to derive greater value from strong, integrated supply chain capabilities, talent becomes a critical driver of success. Managers must shift from focusing on functional expertise and cost-minimization to boundary-spanning solutions that prioritize total cost of ownership. This requires general management skills and business acumen to supplement supply chain expertise.

The expanded role of supply chain managers comes at a time when finding qualified candidates with the appropriate mix of supply chain skills, general management aptitude, and relevant industry knowledge is a growing challenge. Works by McCrea (2012), Gartner (Klappich, Stiffler, & Tohamy, 2011) and Massachusetts Institute of Technology (Cottrill, 2010) allude to a limited pool of experienced candidates with the appropriate skills for the dynamic and changing SCM profession.

This talent shortage is exacerbated by an increased demand for supply chain professionals. The growing recognition among C-level executives that SCM is essential to organizational success is creating competition for top talent. Also, the U.S. Bureau of Labor Statistics (2012) reports that "Employment of logisticians is expected to grow 26 percent from 2010 to 2020, faster than the average for all occupations. Employment growth will be driven by the important role of logistics in an increasingly global economy."

A three-pronged talent management process is needed to thrive in a talent-constrained market. First, organizations must focus on the acquisition of supply chain talent. Position responsibilities, important candidate capabilities, and talent requirement forecasts must be evaluated. Strong recruiting efforts, based on key talent requirements, are needed to attract qualified supply chain candidates. And hiring processes must thoroughly vet candidates in short order to ensure that the best available talent is acquired.

Next, supply chain talent must be developed. A strong onboarding program rapidly assimilates supply chain talent into their roles so that they become productive and satisfied team players. Development continues with ongoing training and education to help managers keep pace with this rapidly changing field, contribute to organizational success, and prepare for future roles. Also, customized individual development plans, mentoring relationships, and high potential programs enhance learning, professional growth, and leadership ability of supply chain managers.

The third talent development component is a strategic talent advancement process. It focuses on creating the bench strength and institutional supply chain knowledge that underpin future success. Well-designed and logical career paths build expertise and leadership, driving professional growth and guiding individuals toward senior supply chain roles. These concerted efforts to provide development and advancement opportunities also mitigate turnover and its related costs. Finally, a forward-focused talent management program used retention planning to build the organization's promotion-ready pool of supply chain managers.

When properly funded and executed, the talent management process helps organizations build high-performance teams with the skills and capabilities to fulfill the expanded supply chain missions being initiated by C-level executives. In contrast, a failure to focus on talent development will reduce SCM bench strength and accelerate turnover of key talent, leaving the organization deficient in the skills needed to leverage SCM to its fullest potential (Gibson, Williams, Gofnett, and Cook, 2013).

Driver 4: Omnichannel Customers

The customer is king (or queen) adage is alive and well in the twenty-first century, thanks to technology. Today, customer demand emanates anywhere and anytime for just about everything. Organizations, particularly retailers, cannot afford to ignore this opportunity. The near-term outlook for U.S. e-commerce sales is strong, with digital sales expected to increase at a 14 percent compound annual growth rate from 2012 to 2017, with sales growing to $434 billion (eMarketer, 2013). This growth will shift additional sales from traditional brick-and-mortar stores, boosting e-commerce from 8 percent to 10 percent of U.S. retail sales. Similar growth is predicted for Europe, led by the U.K. at 15 percent, according to Forrester Research (Indvik, 2013).

With consumers able to shop via mobile devices, home computers, phone, mail, and in-store, organizations must quickly establish effective fulfillment responses. They must develop seamless capabilities to fulfill demand from anywhere in their system to support omnichannel customers. This requires development and synchronization of fulfillment infrastructure, technology, and processes.

Multiple fulfillment infrastructure options exist for omnichannel retailers. Primary setups include dedicated facilities that focus on individual channels with independent inventories, integrated facilities that serve all channels with a common pool of inventory, or a hybrid of the two (Gibson & Defee, 2012). Other options include fulfillment from brick-and-mortar stores, supplier facilities, or logistics service provider facilities. Key fulfillment infrastructure selection considerations include product characteristics, demand volume and geographic dispersion, SKU commonality, and so on. It is not unusual to find retailers using multiple infrastructures to support a mix of fulfillment options.

Accurate capture and timely fulfillment of omnichannel orders via this infrastructure depends on reliable technology. Many of the software tools discussed in Chapter 4 support fulfillment execution, but the future game-changer may be Distributed Order Management (DOM). This software supports fulfillment regardless of order origin: computer, retail store, kiosk, or mobile phone. After order capture, the DOM software determines the best fulfillment facility based on inventory availability, delivery cost, capacity, customer requirements, and other factors. When implemented and executed effectively, DOM helps retailers meet service requirements through the most efficient means possible.

Omnichannel execution requires flexible processes. Appropriate picking, packing, and shipping methods must be established for each type of order: individual e-commerce transactions, retail store orders, and wholesale orders. Optimal flow paths between these operations must also be determined (Fortna, 2013). The subsequent delivery processes must support variable speed to market with order cycle time determined by individual customer preference. Some organizations are ramping up capabilities for same-day fulfillment and delivery in major markets with the expectation that consumers will pay a premium for this service (Evans, 2013).

The bulk of future e-commerce growth will be captured by organizations that creatively align fulfillment infrastructure, technology, and processes. Investing in these capabilities will be essential to any organization that expects to compete companies such as Amazon.com and its global network of 89 distribution centers that continues to grow (Kucera, 2013).

Driver 5: Supply Shifting

High-profile announcements of U.S. companies shifting sourcing and production away from China and other Asian countries have been in the news for a few years. In 2009, Rubin cited oil cost as a major reason for future contraction of offshore sourcing and predicted the end of globalization as we currently know it (Harrington, 2012). Rising wage rates in China and other traditional low-cost country sourcing locations also became a concern to organizations. Add in supply chain risk and a lack of responsiveness, and offshoring lost some of its luster. Soon, talk turned to action, and the pendulum began to swing from offshoring to nearshoring and onshoring.

Though Whirlpool, General Electric, and Ford have brought some production back to the United States, there is no indication that full-scale domestic manufacturing will return. As activity shifts from the Far East to the Americas, key sites will include Mexico, Chile, Brazil, Costa Rica, and other Latin American countries. Potential benefits of nearshoring include lower freight costs, less in-transit inventory, and improved speed to market according to a recent study (AlixPartners, 2012). Many organizations are finding that lower supply chain costs plus greater agility more than offset the higher wage rates in the Americas for a lower total cost of ownership.

Despite the bold predictions that millions of manufacturing jobs will be reshored, the global supply chain will continue to operate. Organizations will continue to offshore manufacturing activity but not at the pace or volume as in the past. Research suggests that there will be more production shifting activity between low-cost countries, from China and India to Viet Nam, Indonesia, and the Philippines, for example. The key to successful production shifting is to properly evaluate the capacity, efficiency, and supply chain capabilities of potential locations (Booth, 2013). For some products such as shoes, near-term production shifting options may be limited because the needed infrastructure is largely concentrated in China.

The future implication for supply chain managers is to take part in the strategic conversation regarding sourcing and manufacturing locations. During the offshoring boom, there was too much attention paid to labor costs and not enough consideration of operational details and supply chain costs, according to Professor Marshall Fisher (Knowledge@Wharton, 2013). As the strategic discussion turns toward onshoring, nearshoring, and reshoring, supply chain managers must actively engage in the conversation. They must bring fact-based logic about transportation costs, inventory costs, agility implications, location-specific risks, and availability of infrastructure, among other key considerations to the strategic decision-making process.

Driver 6: Automation

An important consideration in the "-shoring" debate and supply shifting is the availability automation. The ability to automate high labor content production activities reduces the labor cost share of total cost. It also minimizes the impact of higher wages in North America versus low-cost countries. For example, the availability of automation encouraged Lenovo, a Chinese technology company to produce personal computers in Whitsett, North Carolina (Booth, 2013).

Automation brings a number of benefits to the supply chain. In addition to increasing labor productivity and reducing the labor cost component, manufacturing automation mitigates the effect of labor shortages, eliminates repetitive tasks, increases worker safety, improves product quality, and helps cut manufacturing lead time (O'Sullivan, 2009). For these reasons, companies that do not automate are likely to find themselves at a competitive disadvantage. Looking forward, innovations such as 3D printers; integration of information technology; and less expensive, more flexible robots will make automation more agile and improve return on investment (ROI) (Pinto, 2013).

Automation is not only beneficial for manufacturing. Following the lead of European companies, there has been an uptick in distribution center automation in North America. The aging workforce, combined with the cost and physical challenges of manually selecting store orders, led Kroger and Sobeys to invest millions of dollars in highly automated distribution centers. Employees typically touch a product only on the receiving and shipping docks. Other high-volume retailers, intrigued by the potential gains in order fulfillment efficiency, accuracy, and safety are actively evaluating the business case for similar investments (Gibson & Defee, 2012).

Like any other major investment, automation requires a thorough investigation of the issues, alternatives, benefits, and costs to determine whether automation is appropriate and financially feasible. Supply chain managers need to analyze new automation options, potential supply chain labor shortages, and cost drivers such as mandatory health care coverage. These types of developments can tip the organization's decision toward increased use of automation in their manufacturing and fulfillment operations.

Driver 7: Sustainability

The concepts of environmentalism and sustainability are not new to the supply chain, but they do hold a dual threat/opportunity in the twenty-first century. From the threat perspective, organizations cannot ignore their environmental impact because buyers and consumers increasingly consider the issue during supplier selection (Engel, 2011). Sales will be lost if an organization cannot articulate a strategy and demonstrate improvements in performance. From an opportunity perspective, there are financial benefits of focusing on sustainability. Put another way, organizations can generate green (money) from being green.

Doing more with less is certainly one way to profitably focus on sustainability. Over the last 5 years, Wal-Mart's efforts to redesign and reduce packaging, improve trailer capacity utilization, and drive more efficiently help the company deliver 361 million more cases while driving 287 million fewer miles. Hormel's switch to ultra-lightweight trucks to increase product weight per trailer, reduced fuel consumption by 9,800 gallons and carbon emissions by 100 metric tons (Mayer, 2013). In both cases, the companies cut spending on fuel, used fewer natural resources, and lowered their environmental impact. Further transportation opportunities exist from the use of natural gas engines that use cheaper, more plentiful fuel that burns cleaner.

Turning a reverse supply chain into a profit center can be both financially rewarding and environmentally responsible. Rather than viewing returns and end-of-life cycle products as a costly nuisance that deserves little attention, innovative organizations are setting up supply chain processes and priorities to mitigate reverse costs. Though the simplest option is to dump product into a landfill, it should be the last resort disposition strategy because it sacrifices any remaining product value. Instead, a disposition process using all potential strategies—refurbish, auction, redistribute, recycle, donate, and so on—should be implemented based on financial analysis, regulatory compliance, and brand considerations. Like the forward supply chain, reverse processes should also be monitored through appropriate metrics (Agrawal, 2012).

The need to develop more sustainable supply chains is not only the right thing to do socially and financially, but it is also a growing regulatory compliance issue. Reverse logistics processes must ensure proper end-of-life management according to the Extended Product Responsibility laws in 24 U.S. states and similar regulations around the world. These processes are often complex and will require more time and effort from supply chain managers than they might estimate (Rogers, Lembke, & Benadino, 2013).

Driver 8: Network Design

Collectively, Drivers 4–7 will have a major impact on supply chain network design. Anytime, anywhere sales channels will require new fulfillment facilities or repurposing of existing operations. Changes in manufacturing locations and automation levels may make current sourcing locations, distribution hubs, and delivery routes obsolete or suboptimal. And sustainability goals may drive development of market-facing facilities and reverse networks. Organizations will need to reassess their current designs and make modifications to accommodate these new strategies.

Typically, network design studies occur infrequently due to their cost, complexity, and time requirements. However, the changes listed previously, as well as volatility of demand, mergers, acquisitions, or divestitures should trigger more frequent network reevaluation studies. In unusual situations, organizations evaluate their network design several times per year (Watson, Lewis, Cacioppi, & Jayaraman, 2012). The potential benefit of all this effort is supply chain cost-savings of 12 percent to 20 percent and service level improvements (Bragg, Stone, & Van Geersdaele, 2011).

To gain the maximum value from a design initiative, it makes sense to use network modeling software. These tools make it possible for supply chain managers to model the existing supply chain, simulate different strategies, and conduct what-if analyses based on different locations, number of facilities, and related factors. Each potential option can be evaluated in terms of cost, benefits, and difficulty of execution (Bragg et al., 2011). After the chosen network is accurately depicted in the model, it becomes easier to conduct future network design evaluations.

Going forward, it will be important for supply chain managers to be open-minded about network designs. They must consider global factors that affect risk and cost, in addition to the typical local considerations, when selecting sites (Boyd, 2012). They will need to think about innovative strategies such as the regional clustering designs used by Zara (Sheffi, 2012). And they should approach footprint decisions in a more nuanced way. Long range, they will want to consider networked enterprise models that use "big data" and analytics to respond quickly and decisively to changing conditions and can also pursue long-term opportunities (Manyika et al., 2012).

Driver 9: Big Data

Big data, a term used to describe information stored in databases, both structured repositories such as transportation management systems and unstructured sources such as call center data, has been dubbed "the next big thing" in SCM by consultants, analysts, and researchers. Their thinking is that if software could search through these vast data stockpiles, it might find connections or hidden patterns that could then be parlayed into supply chain improvements (Cooke, 2013). Many of the desirable supply chain capabilities discussed throughout this book, such as inventory visibility, transportation planning, and demand sensing, could be enhanced through big data systems that help managers rapidly convert data into better decisions that save money and optimize operations (Sciacca, 2013).

The unique big data opportunities lay in the raw supply chain data that has not yet been scrubbed clean or organized. Cecere (2012) suggests that this unstructured data from customer interactions, social media, and other external sources will be very important to supply chain managers. Their supply chain processes will need to use the sensing and pattern recognition from big data technologies to quickly listen and learn from previously unused data. This will potentially help managers to better sense true channel demand and respond accordingly. For example, Tesco has developed an analytics model that compares historical buying data and expected temperature trends. The predictive model helps the company accurately predict what consumers in a specific area will want to purchase and then supply locations accordingly. Using this model, Tesco has reduced out of stocks and excess inventory (Master, 2013).

Taking advantage of big data requires a logical process and strong analytics infrastructure. First, data must be collected in real-time. Next, the data must be processed as it flows via calculations, transformation, and augmentation. Finally, the data is explored using ad hoc queries and visualized via dashboards (Kalakota, 2011). The working principle behind big data tools is to move the query to the data to be processed, not the data to the query processor. Every step of the process means the database is changing and needs to be reloaded. Hence, the analytic models are large and require very large amounts of memory to operate. A strong computing layer, file systems and databases, and analytics packages are needed to evaluate the data in terms of the business issue (Floyer, 2013).

Though big data systems offer great promise, supply chain managers must separate the reality from the hype. It is important to focus on quality of data versus quantity of data, articulate data initiatives that will generate the greatest results, and leverage cloud-based storage and analytics rather than build internal solutions (Woods, 2013). Still, with 80 percent of supply chain data coming from external partners versus 20 percent from internal data sets, there is a tremendous opportunity for big data capabilities to create a unified view of the supply chain and one cross-chain version of the truth (Pemmaraju, 2011).

Driver 10: Government Oversight

There is a strong sentiment among supply chain managers that the government is not looking out for their best interests. In a recent Material Handling & Logistics survey, participants cited an increasing level of government intervention as the second biggest challenge behind cost issues (Blanchard, 2013). They believe that ill-defined mandates, onerous regulation, and overly aggressive enforcement will negatively affect fulfillment operations.

Managers across the supply chain must analyze how regulatory changes affect their operations and take steps to limit the negative effect on cost and customer service. For example, U.S. governmental regulation of truck driver hours of service portends additional expense to fulfill demand. Carriers complain of lost productivity and the need for more resources to move the same amount of freight they handle now. Shippers say the new rules reduce vehicle miles driven and force them to reconfigure manufacturing and distribution networks (Solomon, 2013). Proactive and creative process redesigns will be needed to minimize the negative impact.

The regulatory challenge is not limited to U.S. governmental activity. For example, products sold into any European Union country must not contain an expanding list of chemicals deemed toxic by the Union's Registration, Evaluation, Authorization and Restriction of Chemicals, and its Restriction of Hazardous Substances Directive (Westervelt, 2012). Also, mandates and directives related to sustainability, safety, labor protection, and ethical trade are on the rise. Organizations must develop mechanisms to monitor the changing regulatory landscape and determine appropriate supply chain responses. Some are turning to software to assist with the challenge of product ingredient compliance, whereas others seek the assistance of knowledgeable logistics service providers and consultants that are familiar with regional requirements.

At the same time, managers are concerned about inadequate infrastructure investment by the government. The concern is highlighted by a recent report card rating of D+ by the American Society of Engineers and its conclusion that spending levels are falling short by $1.6 trillion (Halsey, 2013). A lack of transportation infrastructure maintenance and expansion is stoking concern that congestion will result in lost productivity, delivery delays, and higher supply chain costs. Supply chain managers will need to consider alternate delivery flows, patterns, and timing to minimize the impact of congestion.

Chapter Summary

A look back to the beginning of this book reveals why organizations must vigilantly monitor these ten driving forces and the supply chain landscape for other emerging issues. Chapter 1 talked about developing an efficient and agile supply chain capabilities to respond to dynamic market requirements is the difference between great success and utter chaos. The driving forces of change discussed in this chapter are certainly dynamic and many of them are aligned with market requirements. Hence, organizations must manage them well to achieve supply chain success and competitive advantage.

As Stock (2013) indicates, there will be technology and a tidal wave of supply chain data to assist managers with these driving forces, but it won't be easy. They will still be challenged to control risk, harness data, and overcome many obstacles. But if any group can do it, it is the creative, dedicated, and energetic individuals that work in the SCM profession. Good luck with the journey!

References

Agrawal, A. (2012) Turn your reverse supply chain into a profit center. *CSCMP's Supply Chain Quarterly,* 6(1), 46–53.

AlixPartners. (2012) Mexico continues to be top choice for manufacturing "near-shoring," says AlixPartners study. Retrieved September 11, 2013, from http://www.alixpartners.com/en/MediaCenter/PressReleaseArchive/tabid/821/articleType/ArticleView/articleId/238/Mexico-Continues-To-Be-Top-Choice-for-Manufacturing-Near-Shoring-Says-AlixPartners-Study.aspx.

Autry, C., Goldsby, T., and Bell, J. E. (2012) *Global macrotrends and their impact on supply chain management,* Upper Saddle River, NJ: FT Press.

Blanchard, D. (2013) Logistically speaking: Whose supply chain is it, yours or the government's? *MH&L News.* Retrieved September 9, 2013, from http://mhlnews.com/global-supply-chain/logistically-speaking-whose-supply-chain-it-yours-or-governments.

Booth, T. (2013) Here, there, and everywhere. Retrieved September 9, 2013, from http://www.economist.com/news/special-report/21569572-after-decades-sending-work-across-world-companies-are-rethinking-their-offshoring.

Boyd, J. (2012) Think globally, site locally. *Supply Chain Quarterly,* August, 24–25.

Bragg, S., Stone, R., Van Geersdaele, J. (2011) 7 signs your supply chain needs a redesign. *CSCMP's Supply Chain Quarterly,* 5(3), 47–52.

Burnson, P. (2013a) Deloitte survey: Executives face growing threats to logistics management. *Logistics Management.* Retrieved September 9, 2013, from http://www.logisticsmgmt.com/article/deloitte_survey_executives_face_growing_threats_to_logistics_management.

Burnson, P. (2013b) PwC's 16th annual global CEO survey reveals continued focus on supply chain. *Logistics Management.* Retrieved September 8, 2013, from http://www.logisticsmgmt.com/article/pwcs_16th_annual_global_ceo_survey_reveals_continued_focus_on_supply_chain.

Cecere, L. (2013) Five supply chain opportunities in big data and predictive analytics. *DataInformed.* Retrieved September 10, 2013, from http://data-informed.com/five-supply-chain-opportunities-in-big-data-and-predictive-analytics/#sthash.K6m1Cf0D.dpuf.

Cooke, J. (2013) Shippers show big interest in big data. *DC Velocity.* Retrieved September 10, 2013, from http://www.dcvelocity.com/articles/20130715-shippers-show-big-interest-in-big-data/.

Cottrill, K. (2010, Fall) Are you prepared for the supply chain talent crisis? MIT CTL White paper. Retrieved February 16, 2013, from http://ctl.mit.edu/sites/default/files/library/public/Talent_FNL_a.pdf.

Cudahy, G. C., George, M. O., Godfrey, G. R., and Rollman, M. J. (2012) Preparing for the unpredictable. *Outlook: The Online Journal of High-Performance Business.* Retrieved August 8, 2013, from http://www.accenture.com/us-en/outlook/Pages/outlook-journal-2012-preparing-for-unpredictable.aspx.

eMarketer (2013) Retail ecommerce set to keep a strong pace through 2017. Retrieved September 10, 2013, from http://www.emarketer.com/Article/Retail-Ecommerce-Set-Keep-Strong-Pace-Through-2017/1009836h-2017/1009836 – KDZBYQ1AXt1DyviB.99.

Engel, B. (2011) 10 best practices you should be doing now. *CSCMP's Supply Chain Quarterly,* 5(1): 48–53.

Ernst & Young. (2011) *Operational agility: From supply chain to integrated value chain.* Retrieved March 20, 2013, from http://www.ey.com/GL/en/Issues/Driving-growth/Growing-Beyond-Operational-agility.

Evans, K. (2013) The rise of same-day deliveries, thanks in part to Amazon. *Internet Retailer.* Retrieved September 10, 2013, from http://www.internetretailer.com/2013/08/29/rise-same-day-deliveries-thanks-part-amazon.

Floyer, D. (2013) Enterprise big-data. *Wikibon*. Retrieved September 10, 2013, from http://wikibon.org/wiki/v/Enterprise_Big-data#Components_of_Big-data_Processing.

Fortna. (2013) *5 steps to designing omni-channel fulfillment operations.* Retrieved September 10, 2013, from http://www.fortna.com/whitepapers/designing-omnichannel-fulfillment-operations-en.pdf.

Gibson, B., and Defee, C. (2012) *The state of the retail supply chain: Essential findings of the third annual study.* Retrieved September 10, 2013, from http://www.rila.org/news/industry/BenchmarkingResearch/Third%20Annual%20SRSC%20Report%20-%20Final.pdf.

Gibson B., Williams, Z., Goffnett, S., and Cook, R. (2013) *SCM talent development: The advance process.* Oak Brook, IL: Council of Supply Chain Management Professionals.

Halsey, A. (2013, March 19) U.S. infrastructure gets D+ in annual report. *The Washington Post.* Retrieved September 9, 2013, from http://articles.washingtonpost.com/2013-03-19/local/37846681_1_civil-engineers-infrastructure-report-card.

Harrington, L. (2012) Nearshoring Latin America: A closer look. *Inbound Logistics.* Retrieved September 11, 2013, from http://www.inboundlogistics.com/cms/article/nearshoring-latin-america-a-closer-look/.

Indvik, L. (2013) Forrester: U.S. online retail sales to hit $370 billion by 2017. *Mashable.* Retrieved September 10, 2013, from http://mashable.com/2013/03/12/forrester-u-s-ecommerce-forecast-2017/.

Kalakota, R. (2011) What is a hadoop? Explaining bid data to the C-Suite. *Practical analytics.* Retrieved September 10, 2013, from http://practicalanalytics.wordpress.com/2011/11/06/explaining-hadoop-to-management-whats-the-big-data-deal/.

Klappich, D., Stiffler, D., and Tohamy, N. (2011) Talent and high performing supply chains. *Gartner Webinars.*

Knowledge@Wharton. (2013) Should manufacturing jobs be "re-shored" to the U.S.? Retrieved September 11, 2013, from http://knowledge.wharton.upenn.edu/article.cfm?articleid=3082.

Kucera, D. (2013, August 29) Why Amazon is on a warehouse building spree. *Bloomberg BusinessWeek Technology.* Retrieved September 10, 2013, from http://www.businessweek.com/articles/2013-08-29/why-amazon-is-on-a-warehouse-building-spree.

Lee, W. B. (2013) How to communicate with your board of directors. *Supply Chain Management Review,* 17(2): 32–39.

Manyika, J., Sinclair, J., Dobbs, R., Strube, G., Rassey, L., Mischke, J. . . . and Ramaswamy, S. (2012) *Manufacturing the future: The next era off global growth and innovation.* Retrieved September 10, 2013, from http://www.mckinsey.com/insights/manufacturing/the_future_of_manufacturing.

Master, N. (2013) Tesco improves supply chain with big data, automated data collection. *RFGen*. Retrieved September 10, 2013, from http://www.rfgen.com/blog/bid/285148/Tesco-Improves-Supply-Chain-with-Big-Data-Automated-Data-Collection.

Mayer, M. (2013) Surpassing sustainability. *Refrigerated & Frozen Foods, 23*(5), 30–37.

McCrea, B. (2012, January/February) What hiring managers are looking for today. *Supply Chain Management Review*, January/February Executive Education Special Supplement, S3–S6.

National Restaurant Association. (2013) Supply chain integral to brands' success, CEOs say. Retrieved September 10, 2013, from http://www.restaurant.org/News-Research/News/Supply-chain-integral-to-brands%E2%80%99-success,-CEOs-say.

O'Sullivan, D. (2009) *Industrial automation: Course notes.* Retrieved September 11, 2013, from http://www.nuigalway.ie/staff-sites/david_osullivan/documents/handout.pdf.

Pemmaraju, K. (2011) Five challenges of managing big data in supply chains. *Sand Hill*. Retrieved September 10, 2013, from http://sandhill.com/article/five-challenges-of-managing-big-data-in-supply-chains/.

Pinto, J. (2013) Manufacturing automation futures. *Automation World*. Retrieved September 11, 2013, from http://www.automationworld.com/industry-business/manufacturing-automation-futures.

PwC. (2013) *Global supply chain survey 2013: Next-generation supply chains efficient, fast and tailored.* Retrieved August 28, 2013, from http://www.pwc.com/et_EE/EE/publications/assets/pub/pwc-global-supply-chain-survey-2013.pdf.

Rasmus, R., Godfrey, G., and Richter, G. (2012) Dynamic operations: Driving profitable growth in the age of permanent volatility. *CSCMP Annual Global Conference 2012.*

Rogers, D., Lembke, R., and Benadino, J. (2013) Reverse logistics: A new core competency. *Supply Chain Management Review,* 17(3), 40–47.

Sciacca, C. (2013) Real-time big data is the next big thing for supply chains. *Who said supply chains are boring?* Retrieved September 10, 2013, from http://supplychainsrock.blogspot.com/2013/06/real-time-big-data-is-next-big-thing.html.

Sheffi, Y. (2012) *Logistics clusters: Delivering value and driving growth.* Cambridge, MA: MIT Press.

Solomon, M. (2013, August 2) Appeals court upholds virtually all of government's driver hours of service rules. *DC Velocity*. Retrieved September 9, 2013, from http://www.dcvelocity.com/articles/20130802-appeals-court-upholds-virtually-all-of-governments-driver-hours-of-service-rules/.

Stock, J. (2013) A look back, a look ahead. *CSCMP's Supply Chain Quarterly,* 7(2), 22–26.

U.S. Bureau of Labor Statistics. (2012) Business and financial: Logisticians. *Occupational Outlook Handbook*. Retrieved February 16, 2013, from http://www.bls.gov/ooh/business-and-financial/logisticians.htm.

Watson, M., Lewis, S., Cacioppi, P., and Jayaraman, J. (2012) *Supply chain network design: Applying optimization and analytics to the global supply chain.* Upper Saddle River, NJ: FT Press.

Westervelt, A. (2012, May 10) How international regulations are changing American supply chains. *Forbes*. Retrieved September 10, 2013, from http://www.forbes.com/sites/amywestervelt/2012/05/10/how-international-regulations-are-changing-american-supply-chains/.

Woods, D. (2013, June 27) Why building a distributed data supply chain is more important than big data. *Forbes*. Retrieved September 10, 2013, from http://www.forbes.com/sites/danwoods/2013/06/27/why-building-a-distributed-data-supply-chain-is-more-important-than-big-data/.

Index

Symbols
3PLs (third-party logistics service providers), 33

A
ABC segmentation analysis, 75, 88-89
accessibility
　freight transportation modes, 163
　information, 115
accumulation function (warehousing), 54
accuracy, information, 115
acquisition tradeoffs, 65
activity-based costing (performance measure), 197-199
adaptability
　global supply chains, 149
　performance measures, 182
　strategy, 86-87
　technology, 119
adaptable structures, tempering volatility, 206
advancement process, talent management, 209
agility, 80
　managing volatility, 206
　technology, 118
allocation function (warehousing), 54
Apple, outsourcing utilization, 90
arms-length relationships, 65
assembly process, global supply chain networks, 159-161
assets
　management performance metrics, 192-193
　turnover, 193
assortment function (warehousing), 54
auto-ID (automatic identification), 138
automatic identification (auto-ID), 138
automation, as driving force of change, 211-212

B
balanced scorecards (performance measure), 195-197
barriers to success, 97
　bullwhip effect, 104
　disparate goals, 102-103
　external pressures, 98-99
　functional silos, 103-104
　limited visibility, 105
　network complexity, 98
　process disconnects, 100
　talent gaps, 99-100
　technology deficiencies, 101
　transactional focus, 101-102
batch production, 49
benchmarking performance, 92
Benetton, postponement principle, 77
BI (business intelligence) tools, 135-136
big data, as driving force of change, 214-215
billing systems, 134
bills of lading documents, 170

binary outcomes (performance measures), 179
BTO (build-to-order) production method, 48
build-to-order (BTO) production method, 48
bullwhip effect, as barrier to success, 104
business
   analytics, 140
   channels, global supply chain management, 152
      communication channels, 153-155
      distribution channels, 155
      interactions, 155
      transaction channels, 153
   utilities, 18-19
business intelligence (BI) tools, 135-136
buying situations, 43

C

capabilities, information technology, 117
   adaptability, 119
   agility, 118
   collaboration, 118
   cross-chain visibility, 118
   differentiation, 119
   optimization, 118
   risk management, 119
   speed to market, 118
capacity factors, freight transportation modes, 163
capacity planning, 47
Carriage and Insurance Paid (CIP), 168
Carriage Paid To (CPT), 168
carrier selection, global supply chain networks, 166-167
cash conversion cycle, 31
cash-to-cash cycle, 31
cause-and-effect forecasting, 37
certificate of end use document, 170
certificate of inspection document, 171
certificate of origin document, 170
CFR (Cost and FReight), 169
challenges, information technology, 120-121
change
   driving forces of
      automation, 211-212
      big data, 214-215
      C-level engagement, 207-208
      government oversight, 215-216
      network design, 213-214
      omnichannel customers, 209-210
      permanent volatility, 206-207
      supply shifting, 210-211
      sustainability, 212-213
      talent management, 208-209
   enhancing business response to, 12-13
channels, global supply chain management, 152
   communication channels, 153-155
   distribution channels, 155
   interactions, 155
   transaction channels, 153
channel-spanning performance strategy, 92
characteristics
   good performance measures, 183-187
   information, 115
      accessibility, 115
      accuracy, 115
      relevance, 115
      reliability, 116
      shared view, 116
      timeliness, 115
      transferability, 116
      usability, 116
      value, 116
CIF (Cost Insurance and Freight), 169
CIP (Carriage and Insurance Paid), 168
C-level engagement, as driving force of change, 207-208
cloud computing emergence, 138-139
collaboration
   cross-chain collaboration strategy, 93-94
   technology, 118
collaborative forecasting, 87
Collaborative Planning, Forecasting, and Replenishment (CPFR) model, 40-41, 73

commodities, 43
communications, global supply chain management, 151-155
complexity reduction strategy, 93
configure-to-order (CTO) computer systems (Dell), 14
congestion, intermodal transportation, 172
contingency planning, 95-96
continuous production, 46, 49
contract purchasing, 65
control
   procurement, 45
   transportation, 59-60
Cost and FReight (CFR), 169
cost factors, freight transportation modes, 164
Cost Insurance and Freight (CIF), 169
cost performance metrics, 187-189
cost reporting, 188
cost-to-serve (CTS) model, 62
Council of Supply Chain Management Professionals, 3-5
CPFR (Collaborative Planning, Forecasting, and Replenishment) model, 40-41, 73
CPT (Carriage Paid To), 168
criticals, 43

CRM (customer relationship management), 62-63, 137
cross-chain
   collaboration strategy, 93-94
   metrics, 67
   resources, 10
   visibility, technology, 118
CTS (cost-to-serve) model, 62
cultural factors, global supply chain management, 150-151
currencies, global business challenges, 149
customers
   customer benefits perspective (balanced scorecards), 196
   customer-facing measures (warehouses), 55
   customer relationship management (CRM), 62-63, 137
   customer service performance metrics, 193-194
   deliver process, 51
      transportation, 56-60
      warehousing, 52-55
   demand planning, 38
   driving customer value, 11-12
   global supply chain management, 147
      business channels, 152-155
      cultural and communication factors, 150-151
      financial and investment related factors, 149-150

      legal, political, and environmental factors, 151-152
   omnichannel, as driving force of change, 209-210
   post-sale support, 63
   utility, 18-19
customization principle, 75-76
cycle time (orders) measure, 194

D

damage measurement, 191
DAP (Delivered At Place), 168
data
   big data, 214-215
   POS (point-of-sale), 73
DAT (Delivered At Terminal), 168
days-of-supply metric, 193
DDP (Delivered Duty Paid), 168
deficiencies in technology, as barrier to success, 101
Delivered At Place (DAP), 168
Delivered At Terminal (DAT), 168
Delivered Duty Paid (DDP), 168
deliver process, 51
   order management, 60-62
   transportation, 56-60
   warehousing, 52-55

Dell
  configure-to-order (CTO) computer systems, 14
  individual customization, 90
  segmentation, 89
demand
  demand driven process, 72-73
  dependent demands, 35
  derived demands, 35
  disconnect from supply chains, 100
  global supply chain management, 147
    business channels, 152-155
    cultural and communication factors, 150-151
    financial and investment related factors, 149-150
    legal, political, and environmental factors, 151-152
  matching supply with, 10
  planning, 35
    connecting with supply chain activities, 40
    customer requirements, 38
    forecasting, 36-37
    interfacing between marketplace and manufacturing, 39
    marketing and sales plans, 37-38
    software, 11, 128
  sensing, 86-87

dependent demands, 35
derived demands, 35
dialects, global trade challenges, 151
differentiation, technology, 119
direct stakeholders, 7-8
disparate goals, as barrier to success, 102-103
disruption management capabilities, 14
Distributed Order Management (DOM), 133-134, 210
distribution
  channels, 6, 155
  global supply chain networks, 161-166
  planning software, 129
distributors, 8
DMAIC (Define, Measure, Analyze, Improve, Control), 199-200
DOM (Distributed Order Management), 133-134, 210
driving customer value, 11-12
dyadic metrics, 194

E

EDI (electronic data interchange), 60
effectiveness
  performance measures, 181-182
  strategic profit model, 201

efficiency
  performance measures, 181
  strategic profit model, 200
efficiency principle, 81-82
efficiency transportation metric, 60
electronic data interchange (EDI), 60
elements, order management, 60-62
emerging issues, 23
engineer-to-order (ETO) production method, 48
Enterprise Resource Planning (ERP) systems, 101, 136-137
environmental factors, global supply chain management, 151-152
ERP (Enterprise Resource Planning) systems, 101, 136-137
ETO (engineer-to-order) production method, 48
evaluation principle, 78-79
event management software, 134-135
evolution, supply chain management, 15-18
execution
  applications (supply chain software), 129-134
    DOM (Distributed Order Management), 133-134
    GTM (Global Trade Management), 133

MES (Manufacturing Execution System), 132-133
  TMS (Transportation Management Systems), 131-132
  WMS (Warehouse Management Systems), 131
 procurement, 45
 production process, 47-50
 service process, 50-51
 transportation, 59-60
executive S&OP meetings, 38
export documents, global trading terms, 169
export licenses, 169
Extended Product Responsibility laws, 213
external pressures, as barrier to success, 98-99
EXW (EX Works), 168
EX Works (EXW), 168

F

fabrication shops, 49
facilitating tools (supply chain software), 136-137
facilitators, 9
FAS (Free Alongside Ship), 169
FCA (Free CArrier), 168
fill rate metric, 193
financial
  benefits perspective (balanced scorecards), 197

 factors, global supply chain management, 149-150
 institutions, 9
 outcomes, performance measures, 200-201
 success, facilitating with SCM, 14-15
 tradeoffs, 64
 transactions, 31-32
fixed assets, 192
flow production, 49
flows, 30
  information, 31, 113-114
  monetary, 31-32
  product and related service, 31
FOB (Free On Board), 169
forecasting, 36-37, 87
form utility, 18
forward information flows, 31
framework, supply chain software, 122
  BI (business intelligence) tools, 135-136
  execution applications, 129-134
  facilitating tools, 136-137
  future outlook, 138-140
  planning applications, 125-129
  SCEM applications, 134-135
  SRM software, 137-138
Free Alongside Ship (FAS), 169

Free CArrier (FCA), 168
Free On Board (FOB), 169
freight
  documentation, 59
  manifest documents, 170
  modal capabilities, 161
    accessibility factors, 163
    capacity factors, 163
    cost factors, 164
    reliability factors, 164
    safety factors, 164
    transit times factors, 163
  rates, 58
fulfillment
  order fulfillment, 209-210
  Seven Rights of Fulfillment, 19-20
  tradeoffs, 65
functional areas, global supply chain networks, 156
  carrier selection, 166-167
  distribution/logistics functions, 161-166
  manufacturing/assembly processes, 159-161
  supply management/purchasing, 156-159
functional silo syndrome, as barrier to success, 103-104
future of SCM, driving forces
  automation, 211-212
  big data, 214-215
  C-level engagement, 207-208

government oversight, 215-216
network design, 213-214
omnichannel customers, 209-210
permanent volatility, 206-207
supply shifting, 210-211
sustainability, 212-213
talent management, 208-209

G

generics, 42
geographic scope, network design, 66
global optimization strategy, 87-88
global supply chain management, 23, 147
  business channels, 152
    communication channels, 153-155
    distribution channels, 155
    interactions, 155
    transaction channels, 153
  cultural and communication factors, 150-151
  financial and investment related factors, 149-150
  legal, political, and environmental factors, 151-152
  logistics service providers, 172-173
  network functions, 156
    carrier selection, 166-167
    distribution/logistics functions, 161-166
    manufacturing/ assembly processes, 159-161
    supply management/ purchasing, 156-159
  response to customer demand, 147-149
  route planning, 171-172
  trading terms, 167
    any transportation mode, 168
    export documents, 169
    import documents, 170
    maritime, 169
    transportation documents, 170
global trade management (GTM) systems, 133, 154
goals
  supply chain management, 10
    building network resiliency, 13-14
    drive customer value, 11-12
    efficient fulfillment of demands, 10
    enhancing responsiveness to change, 12-13
    facilitating financial success, 14-15
  tradeoff management, 64
goods and services, procurement, 41
  execution and control, 45
  objectives and organization, 42-43
  supplier evaluation, 46
  supplier selection and negotiation, 43-45
government agencies, 9
government oversight, as driving force of change, 215, 216
GTM (global trade management) systems, 133, 154

H

Hewlett Packard, postponement principle, 77
history, supply chain management, 15-18
holistic performance measures, 194-195

I

import documents, global trading, 170
INCOTERMS (INternational COmmercial TERMS), 167-169
independent demands, 35
indirect material suppliers, 9
information flows, 31
information technology, 91, 111
  capabilities, 117
    adaptability, 119
    agility, 118
    collaboration, 118
    cross-chain visibility, 118
    differentiation, 119
    optimization, 118
    risk management, 119
    speed to market, 118
  challenges, 120-121

characteristics, 115
  accessibility, 115
  accuracy, 115
  relevance, 115
  reliability, 116
  shared view, 116
  timeliness, 115
  transferability, 116
  usability, 116
  value, 116
flows, 113-114
framework, 122-123
needs, 112-113
supply chain software, 123
  BI (business intelligence) tools, 135-136
  execution applications, 129-134
  facilitating tools, 136-137
  future outlook, 138-140
  planning applications, 125-129
  SCEM applications, 134-135
  SRM software, 137-138
innovation and learning perspective (balanced scorecards), 196
innovation (supply chain), managing volatility, 207
integrated networks, 3
integration principle, 81
intermodal transportation, 56, 165, 171
internal measures, warehouses, 55

INternational COmmercial TERMS (INCOTERMS), 167-169
in-transit visibility, transportation execution, 59
inventory
  management, 159
  procurement investment, 46
  turnover, 192
investment factors, global supply chain management, 149-150

J

JIT (Just-In-Time) support, 53
jobbing, 49
job production, 49
job shops, 49
just-in-time delivery, 11
Just-In-Time (JIT) support, 53

K

key participants, 7
  direct stakeholders, 7-8
  facilitators, 9
key performance indicators (KPIs), 55, 79
Kimberly-Clark, creating demand-driven supply chain, 11
KPIs (key performance indicators), 55, 79

L

labor
  automation, 211-212
  management systems, 134
  productivity, 190
  strikes, 13
land bridge transportation, 172
languages, global trade challenges, 151
lead times, order fulfillment, 51
lean logistics strategy, 94-95
lean supply chains, 158
legal factors, global supply chain management, 151-152
limited visibility, as barrier to success, 105
linear representation (supply lines), 3
logistics
  global supply chain networks, 161-166
  LSPs (logistics service providers), 172-173
  management, 6

M

machine shops, 49
maintenance, repair, and operating (MRO) items, 42-43
make process (manufacturing), 46
  capacity planning and scheduling, 47
  production process execution, 47-50
  role of service operations, 46
  service process execution, 50-51

Index 227

make-to-order (MTO) production method, 48
make-to-stock (MTS) production method, 48
manufacturers, 8, 39
manufacturing cells, 50
manufacturing execution system (MES), 132-133
manufacturing process, 46
    capacity planning and scheduling, 47
    global supply chain networks, 159-161
    production process execution, 47-50
    role of service operations, 46
    service process execution, 50-51
maritime, global trading, 169
marketplace, interfacing with manufacturers, 39
market planning, 37-38
mass customization strategy, 90-91
Master Model of Supply Chain Excellence, 122
master planning applications (supply chain software), 127
materials planning, 47
McCain Foods, just-in-time delivery, 11
measurement (performance), 177
    activity-based costing, 197-199

balanced scorecard, 195-197
characteristics of good measures, 183-187
financial outcomes, 200-201
qualitative measures, 178
quantitative measures, 178, 180, 184
role of, 177-178
scope, 194-195
Six Sigma DMAIC, 199-200
trade-off analysis, 180
    focusing the supply chain, 180-182
    understanding the trade-offs, 182-183
types of metrics, 187
    asset management performance, 192-193
    cost performance, 187-189
    customer service performance, 193-194
    productivity, 189-190
    quality performance, 191-192
MES (manufacturing execution system), 132-133
metrics, 187
    asset management performance, 192-193
    cost performance, 187-189
    customer service performance, 193-194

    productivity, 189-190
    quality performance, 191-192
minimum inventory investment, procurement, 46
mobility solutions, 139
modal capabilities, freight transportation, 161
    accessibility factors, 163
    capacity factors, 163
    cost factors, 164
    reliability factors, 164
    safety factors, 164
    transit times factors, 163
modeling software, networks, 213
monetary flows, 31-32
monitoring service quality, transportation execution, 59
MRO (maintenance, repair, and operating) items, 42-43
MTS (make-to-stock) production method, 48

N

natural disasters, 13
nearshoring, 93, 210
negotiation process, supplier selection, 45
network complexity, as barrier to success, 98
network design
    as driving force of change, 213-214
    software, 127

networks
    global supply chain management, 156
        carrier selection, 166-167
        distribution/logistics functions, 161-166
        manufacturing/ assembly processes, 159-161
        supply management/ purchasing, 156-159
    resiliency, 13-14
    structure tradeoffs, 65
    supply chain representation, 3

O

objectives, procurement, 42-43
Ocean Spray, cutting costs and emissions, 10
offshore manufacturing, network complexity, 98
Oliver, Keith, 15
omnichannel customers, as driving force of change, 209-210
one-off production, 49
onshoring, 210
operations planning, 46
    production process execution, 47-50
    service process execution, 50-51
optimization, technology, 118
order cycle time measure, 194
order fill rate, 193

order fulfillment
    deliver process, 51
    order management, 60-62
    transportation, 56-60
    warehousing, 52-55
    key elements, 60-62
    omnichannel customers, 209-210
order-to-cash (OTC) cycle concept, 60-61
organization, procurement, 42-43
origins, supply chain management, 15
OTC (order-to-cash) cycle concept, 60-61
outsourcing strategy, 89-90

P

packaging, global trade, 161
packing lists, 170
participant networks, 7
    direct stakeholders, 7-8
    facilitators, 9
payments, monetary flows, 31-32
perfect order metric, 92
performance measurement, 23, 177
    activity-based costing, 197-199
    balanced scorecard, 195-197
    channel-spanning performance, 92
    characteristics of good measures, 183-187

financial outcomes, 200-201
qualitative measures, 178
quantitative measures, 178, 180, 184
role of, 177-178
scope, 194-195
Six Sigma DMAIC, 199-200
supplier sustainability audits, 96-97
trade-off analysis, 180
    focusing the supply chain, 180-182
    understanding the trade-offs, 182-183
types of metrics, 187
    asset management performance, 192-193
    cost performance, 187-189
    customer service, 193-194
    productivity, 189-190
    quality performance, 191-192
permanent volatility, as driving force of change, 206-207
perspectives, SCM, 4-5
place utility, 18
plan process, 35
    applications (supply chain software), 125-129
        demand planning, 128
        distribution planning, 129
        master planning, 127
        network design, 127

Index 229

procurement planning, 128
production planning, 128
S&OP (sales and operations planning), 127
demand planning, 35
  connecting with supply chain activities, 40
  customer requirements, 38
  forecasting, 36, 37
  interfacing between marketplace and manufacturing, 39
  marketing and sales plans, 37-38
point-of-sale (POS) data, 11, 73
political factors, global supply chain management, 151-152
POS (point-of-sale) data, 11, 73
possession utility, 18
postponement principle, 76-77
post-sale support, 63
predisruption steps, 14
premium service requirements, 19
pre-S&OP meetings, 38
price, supplier selection, 44
primary processes. *See* processes
principles, 71
  agility, 80
  customization, 75-76
  defined, 72

demand driven process, 72-73
efficiency, 81-82
evaluation, 78-79
integration, 81
postponement, 76-77
resiliency, 83
segmentation, 74-75
sustainability, 84-85
total cost focus, 73-74
tradeoff management, 66-67
visibility, 77-78
processes, 22, 29, 34
  deliver, 51
    order management, 60-62
    transportation, 56-60
    warehousing, 52-55
  disconnects, as barrier to success, 100
  make, 46
    capacity planning and scheduling, 47
    manufacturing and service operations, 46
    production process execution, 47-50
    service process execution, 50-51
  plan, demand planning, 35-40
  return, 62
    CRM (customer relationship management), 62-63
    post-sale support, 63
  source, procurement, 41-46

processing speed, technology integration, 91
process standardization, 148
procurement, 41
  execution and control, 45
  objectives and organization, 42-43
  planning software, 128
  supplier evaluation, 46
  supplier selection and negotiation, 43-45
product and related service flows, 31
product customization, 148
product-handling functions, warehousing, 55
production planning software, 128
production processes, 47-50
productivity metrics, 189-190
product safety, global trade transportation, 172
project layouts, 49
project manufacturing, 48
pull model, inventory management, 159
purchased items, 42-43
purchasing function, global supply chain networks, 156-159
push model, inventory management, 159

230   Index

Q

qualitative forecasting methods, 36
qualitative performance measures, 178
quality
  improvement, procurement, 46
  performance metrics, 191-192
  service, transportation execution, 59
  suppliers, 43
quantitative forecasting methods, 37
quantitative performance measures, 178, 180, 184
quantity purchase discounts, 53
quantity utility, 19

R

range forecasts, 87
rebuys, 43
reliability
  freight transportation modes, 164
  information, 116
  suppliers, 44
reporting costs, 188
resiliency principle, 83
responsive strategy, 86-87
retailer point-of-sale data, 11
retailers, 8
return process, 62
  CRM (customer relationship management), 62-63
  post-sale support, 63

reverse information flow, 31
risk
  assessment, 14
  identification, 14
  management, 95-96, 119
  reduction, 14
route planning, global trades, 171-172
routine decision making, 113

S

safety factors, freight transportation modes, 164
sales and operations planning (S&OP), 37-38, 127
SCEM (supply chain event management) applications (software), 134-135
SCM (supply chain management)
  concepts, 2-6
    distribution channels, 6
    logistics management, 6
    supply management, 6
    value chain, 6
  emerging issues, 23
  evolution, 15-18
  flows, 30
    information, 31
    monetary, 31-32
    product and related service, 31

goals perspective (balanced scorecards), 196
global supply chain management, 23, 147
  business channels, 152-155
  cultural and communication factors, 150-151
  financial and investment related factors, 149-150
  legal, political, and environmental factors, 151-152
  logistics service providers, 172-173
  network functions, 156-167
  response to customer demand, 147-149
  route planning, 171-172
  trading terms, 167-170
information technology, 111
  capabilities, 117-119
  challenges, 120-121
  characteristics, 115-117
  flows, 113-114
  framework, 122-123
  needs, 112-113
  supply chain software, 123-140
key participants, 7
  direct stakeholders, 7-8
  facilitators, 9

performance assessment, 23
perspectives, 4-5
processes, 22, 29, 34
   deliver, 51-62
   make, 46-51
   plan, 35-40
   return, 62-63
   source, 41-46
purpose and goals, 10
   building network resiliency, 13-14
   drive customer value, 11-12
   efficient fulfillment of demands, 10
   enhancing responsiveness to change, 12-13
   facilitating financial success, 14-15
role of technology, 23
strategic principles, 71
   agility, 80
   customization, 75-76
   defined, 72
   demand driven process, 72-73
   efficiency, 81-82
   evaluation, 78-79
   integration, 81
   postponement, 76-77
   resiliency, 83
   segmentation, 74-75
   sustainability, 84-85
   total cost focus, 73-74
   visibility, 77-78
strategies, 23, 85
   ABC segmentation analysis, 88-89
   barriers, 97-105
   channel-spanning performance, 92
   complexity reduction, 93
   cross-chain collaboration, 93-94
   demand sensing, 86-87
   global optimization, 87-88
   lean logistics, 94-95
   mass customization, 90-91
   outsourcing, 89-90
   responsive, 86-87
   risk management and contingency planning, 95-96
   supplier sustainability audits, 96-97
   technology integration, 91-92
supply chain structure, 32-33
tradeoff management, 64
   acquisition tradeoffs, 65
   financial tradeoffs, 64
   fulfillment tradeoffs, 65
   goals, 64
   network structure tradeoffs, 65
   principles, 66-67
value propositions, 18
   customer utility, 18-19
   Seven Rights of Fulfillment, 19-20
   shareholder value, 20-22
scope, supply chain performance, 194-195
scorecards, 186
   balanced scorecard measure, 195-197
   suppliers, 92
SCOR (Supply Chain Operations Reference) Model, 34
segmentation principle, 74-75, 88-89
self-owned facilities, 66
service operations, 46
   capacity planning and scheduling, 47
   production process execution, 47-50
   service process execution, 50-51
service quality, 59
Seven Rights of Fulfillment, 19-20
shared view, information, 116
shareholder value, 20-22
shipment preparation, transportation execution, 59
shippers export declaration document, 170
shippers letter of instruction, 170
silos, as barrier to success, 103-104
Six Sigma DMAIC (performance measure), 199-200
slotting software tools, 134
soft performance measures, 178

software
- network modeling, 213
- supply chain, 123
  - *BI (business intelligence) tools, 135-136*
  - *execution applications, 129-134*
  - *facilitating tools, 136-137*
  - *future outlook, 138-140*
  - *planning applications, 125-129*
  - *SCEM applications, 134-135*
  - *SRM software, 137-138*

S&OP (sales and operations planning), 37-38, 127

sortation function (warehousing), 54

source plan, procurement, 41
- execution and control, 45
- objectives and organization, 42-43
- supplier evaluation, 46
- supplier selection and negotiation, 43-45

speed to market, technology, 118

spot market purchases, 65

SRM (supplier relationship management) software, 137-138

stakeholders, 7-8

standard service requirements, 19

stockouts, 73

strategic decision making, 112

strategic principles, 71
- agility, 80
- customization, 75-76
- defined, 72
- demand driven process, 72-73
- efficiency, 81-82
- evaluation, 78-79
- integration, 81
- postponement, 76-77
- resiliency, 83
- segmentation, 74-75
- sustainability, 84-85
- total cost focus, 73-74
- visibility, 77-78

Strategic Profit Model, 20-22, 200-201

strategies, 23, 85
- ABC segmentation analysis, 88-89
- barriers, 97
  - *bullwhip effect, 104*
  - *disparate goals, 102-103*
  - *external pressures, 98-99*
  - *functional silos, 103-104*
  - *limited visibility, 105*
  - *network complexity, 98*
  - *process disconnects, 100*
  - *talent gaps, 99-100*
  - *technology deficiencies, 101*
  - *transactional focus, 101-102*
- channel-spanning performance, 92
- complexity reduction, 93
- cross-chain collaboration, 93-94
- demand sensing, 86-87
- global optimization, 87-88
- lean logistics, 94-95
- mass customization, 90-91
- outsourcing, 89-90
- responsive, 86-87
- risk management and contingency planning, 95-96
- supplier sustainability audits, 96-97
- technology integration, 91-92

structure (supply chains), 32-33

supplier-facing metrics, 92

supplier relationship management (SRM) software, 137-138

suppliers, 8
- consolidation, 93
- procurement evaluation, 46
- scorecards, 92
- selection and negotiation, 43-45
- sustainability audits, 96-97

supply chain event management (SCEM) applications (software), 134-135

supply chain management. *See* SCM

Supply Chain Operations Reference (SCOR) Model, 34

supply chains
  connecting with demand management, 40
  disconnect from demand, 100
  information characteristics, 115
    accessibility, 115
    accuracy, 115
    relevance, 115
    reliability, 116
    shared view, 116
    timeliness, 115
    transferability, 116
    usability, 116
    value, 116
  information flows, 113-114
  information needs, 112-113
  structure, 32-33
Supply Chain Top 25 rankings (Gartner, Inc.), 11
supply shifting, as driving force of change, 210-211
support functions, warehousing, 55
sustainability, 84-85
  as driving force of change, 212-213
  supplier sustainability audits, 96-97

T

tactical planning, 113
talent gaps, as barrier to success, 99-100
talent management, as driving force of change, 208-209
TCO (Total Cost of Ownership), 44, 88
technology, 23, 111-112
  capabilities, 117
    adaptability, 119
    agility, 118
    collaboration, 118
    cross-chain visibility, 118
    differentiation, 119
    optimization, 118
    risk management, 119
    speed to market, 118
  challenges, 120-121
  deficiencies, as barrier to success, 101
  framework, 122-123
  information characteristics, 115
    accessibility, 115
    accuracy, 115
    relevance, 115
    reliability, 116
    shared view, 116
    timeliness, 115
    transferability, 116
    usability, 116
    value, 116
  information flows, 113-114
  information needs, 112-113
  role in supply chain tradeoff, 66
  supply chain software, 123
    BI (business intelligence) tools, 135-136
    execution applications, 129-134
    facilitating tools, 136-137
    future outlook, 138-140
    planning applications, 125-129
    SCEM applications, 134-135
    SRM software, 137-138
technology integration strategy, 91-92
terminology, 2
  distribution channels, 6
  global trading, 167
    any transportation mode, 168
    export documents, 169
    import documents, 170
    maritime, 169
    transportation documents, 170
  logistics, 6
  supply management, 6
  value chain, 6
third-party logistics service provider (3PL), 33
Tier 1 suppliers, 8
Tier 2 suppliers, 8
Tier 3 suppliers, 8
time-series forecasting, 37
time utility, 18
TLC (Total Landed Cost) comparative analysis, 88
TMS (Transportation Management Systems), 131-132
total cost focus, 73-74
Total Cost of Ownership (TCO), 44, 88

Total Landed Cost (TLC) comparative analysis, 88
trade-offs
　analysis, performance measures, 180
　　focusing the supply chain, 180-182
　　understanding the trade-offs, 182-183
　management, 64
　　acquisition tradeoffs, 65
　　financial tradeoffs, 64
　　fulfillment tradeoffs, 65
　　goals, 64
　　network structure tradeoffs, 65
　　principles, 66-67
trading terms, global supply chain management, 167
　any transportation mode, 168
　export documents, 169
　import documents, 170
　maritime, 169
　transportation documents, 170
transactional focus, as barrier to success, 101-102
transaction channels, global supply chain management, 153
transaction processing, 113
transferability, information, 116
transfer of ownership, global trade, 153

transformation process, global trade, 156
transit times factors, freight transportation modes, 163
transportation
　deliver process, 56-60
　documents, global trading, 170
　freight modal capabilities, 161
　　accessibility factors, 163
　　capacity factors, 163
　　cost factors, 164
　　intermodal transportation, 165
　　reliability factors, 164
　　safety factors, 164
　　transit times factors, 163
Transportation Management Systems (TMS), 131-132

U

upstream metrics, 92
usability, information, 116
utilities, 18-19

V

value
　chain, 6
　information, 116
　propositions, 18
　　customer utility, 18-19
　　Seven Rights of Fulfillment, 19-20
　　shareholder value, 20-22

Vendor Managed Inventory (VMI), 73
VICS (Voluntary Inter-industry Commerce Standards Committee), 40
visibility
　limitations, as barrier to success, 105
　principle, 77-78
　technology integration, 91
VMI (Vendor Managed Inventory), 73
volatility, as driving force of change, 206-207
Voluntary Inter-industry Commerce Standards Committee (VICS), 40

W

Wal-Mart, sustainability strategy, 97
Warehouse Management Systems (WMS), 131
warehousing, 52-55
　product-handling functions, 55
　support functions, 55
weather disasters, 13
Whirlpool, agility, 80
wholesalers, 8
WMS (Warehouse Management Systems), 131
workcenters, 50
working capital, 192

X–Z

yard management systems, 134